NANOTECHNOLOGICAL ADVANCEMENTS IN THE FOOD INDUSTRY

NANOTECHNOLOGICAL ADVANCEMENTS IN THE FOOD INDUSTRY

Ishrat Majid, PhD

Bababode Adesegun Kehinde, PhD

Sheikh Shafat Hussain, PhD

Editors

AAP | APPLE ACADEMIC PRESS

First edition published 2026

Apple Academic Press Inc.
1265 Goldenrod Circle, NE,
Palm Bay, FL 32905 USA

760 Laurentian Drive, Unit 19,
Burlington, ON L7N 0A4, Canada

CRC Press
2385 NW Executive Center Drive,
Suite 320, Boca Raton FL 33431

4 Park Square, Milton Park,
Abingdon, Oxon, OX14 4RN, UK

Apple Academic Press exclusively co-publishes with CRC Press, an imprint of Taylor & Francis Group, LLC

Library and Archives Canada Cataloguing in Publication

CIP data on file with Canada Library and Archives

Library of Congress Cataloging-in-Publication Data

CIP data on file with US Library of Congress

ISBN: 978-1-77491-974-3 (hbk)
ISBN: 978-1-77491-975-0 (ebk)
DOI: 10.1201/9781774919750

About the Editors

Ishrat Majid, PhD
Scientist, Department of Food Technology at Islamic University of Science and Technology, Jammu and Kashmir, India

Ishrat Majid, PhD, is a Scientist in the Department of Food Technology at Islamic University of Science and Technology, Jammu and Kashmir, India. She has been associated with Lovely Professional University, Punjab, as an Assistant Professor and Visiting Faculty at Amity University, Rajasthan, India. She has attended many international and national conferences, seminars, and workshops within India as well as abroad. She has published more than 30 research and review articles in journals of international repute. She has one edited book published with Springer to her credit and has also contributed 25 book chapters. Dr. Majid serves as an editorial board member and associate/guest editor of many international journals. She is a peer reviewer with reputed international journals of food science and technology affiliated with Taylor and Francis, Elsevier, Wiley, and Springer etc. She is also working as a member of various associations like International Association for Agricultural Sustainability and AFSTI. She has received a best paper award at the International Conference on Food Properties (iCFP2018) at Sharjah, UAE. Dr. Majid is the recipient of the prestigious award of "WOS-A" from the Department of science and Technology, India. Dr. Majid has been a recipient of international travel grants twice from CSIR & DST, New Delhi, India. She earned her doctoral degree from Sant Longowal Institute of Engineering and Technology, Punjab and master's degree (MTech) from Jamia Hamdard, New Delhi, India. She has qualified the National Eligibility Test (NET) and was awarded Senior Research Fellow by CSIR.

Bababode Adesegun Kehinde, PhD
Research Technologist, Penn State University,
State College Pennsylvania, USA

Bababode Adesegun Kehinde, PhD, is currently working as a Research Technologist at Penn State University, State College Pennsylvania, USA. He was awarded his PhD from the University of Kentucky in Lexington, Kentucky, USA.

Sheikh Shafat Hussain, PhD
Researcher, Department of Fisheries,
Government of Jammu & Kashmir, India

Sheikh Shafat Hussain, PhD, is a Researcher in the Department of Fisheries, Government of Jammu & Kashmir, India. He was awarded his PhD in Fisheries Management from Sher-e-Kashmir University of Agricultural Sciences and Technology, Srinagar, Jammu and Kashmir, India.

Contents

Contributors

Anika
Department of Food Science and Technology, Dr. Y. S. Parmar University of Horticulture and Forestry, Nauni, Solan, Himachal Pradesh, India

Aamir Hussain Dar
Department of Food Technology, Islamic University of Science and Technology, Awantipora, Kashmir, India

Kshirod Kumar Dash
Department of Food Processing Technology, Ghani Khan Choudhury Institute of Engineering and Technology, Malda, West Bengal, India

Atul Dhiman
Department of Food Science and Technology, Dr. Y. S. Parmar University of Horticulture and Forestry, Nauni, Solan, Himachal Pradesh, India

Yogesh Gat
Department of Food Engineering and Technology, Institute of Chemical Technology, Mumbai, Maharashtra, India

Supriya Singh Gaur
Department of Food Technology and Nutrition, Lovely Professional University, Phagwara, Punjab, India

Neha Gautam
Department of Basic Sciences, Dr. Y. S. Parmar University of Horticulture and Forestry, Nauni, Solan, Himachal Pradesh, India

Papiha Gawande
Department of Food Engineering and Technology, Institute of Chemical Technology, Marathwada Campus, Jalna, India

Harpreet Kaur
Department of Chemistry, School of Chemical Engineering and Physical Sciences, Lovely Professional University, Phagwara, Punjab, India

Pooja Kesarkar
Department of Food Engineering and Technology, Institute of Chemical Technology, Marathwada Campus, Jalna, India

Ashish Kumar
NCE, Bihar Engineering University, Department of Science and Technology, Government of Bihar, India

Satish Kumar
Department of Food Science and Technology, Dr. Y. S. Parmar University of Horticulture and Forestry, Nauni, Solan, Himachal Pradesh, India

Vikas Kumar
ICAR-Indian Institute of Millets Research, Rajendranagar, Hyderabad, Telangana, India

Nidhi
Department of Chemistry, Lovely Professional University, Phagwara, Punjab, India

Vinay Kumar Pandey
Research and Development Cell, Biotechnology Department, Manav Rachna International Institute of Research and Studies (Deemed to Be University) Faridabad, Haryana, India

Akshay Sharma
Department of Chemistry, Himachal Pradesh University, Shimla, Himachal Pradesh, India

Renuka Sharma
Department of Chemistry, Himachal Pradesh University, Shimla, Himachal Pradesh, India

Aadisha Saini
ICAR-Indian Institute of Millets Research, Rajendranagar, Hyderabad, Telangana, India

Rafeeya Shams
Department of Food Technology and Nutrition, Lovely Professional University, Phagwara, Punjab, India

Chitra Sharma
Department of Chemistry, School of Chemical Engineering and Physical Sciences, Lovely Professional University, Phagwara, Punjab, India

Kartik Sharma
Unit of Innovative Food Packaging and Biomaterials, School of Agro-Industry, Mae Fah Luang University, Chiang Rai, Thailand

Praveen Kumar Sharma
Department of Chemistry, Lovely Professional University, Phagwara, Punjab, India

Jagmohan Singh
Division of Food Science and Technology, Sher-e-Kashmir University of Agricultural Sciences and Technology Jammu, India

Rahul Singh
Department of Bioengineering, Integral University, Lucknow, Uttar Pradesh, India

Shivangi Srivastava
Department of Food Technology, Harcourt Butler Technical University, Nawabganj, Kanpur, Uttar Pradesh, India

Priyanka Suthar
Department of Food Science and Technology, Dr. Y. S. Parmar University of Horticulture and Forestry, Nauni, Solan, Himachal Pradesh, India

Abhinay Thakur
Department of Chemistry, School of Chemical Engineering and Physical Sciences, Lovely Professional University, Phagwara, Punjab, India

Priyanka Thakur
Department of Food Science and Technology, Dr. Y. S. Parmar University of Horticulture and Forestry, Nauni, Solan, Himachal Pradesh, India

Ramesh Chand Thakur
Department of Chemistry, Himachal Pradesh University, Shimla, Himachal Pradesh, India

Anjali Tripathi
School of Health Sciences and Technology, UPES University Dehradun, Dehradun, India

Preface

The current breakthrough in the field of food nanotechnology is an amalgamation of food technology and nanotechnology that is based upon recent advances and innovations in the food industry. It mainly covers synthesis, analysis, characterization, and interaction between food science and nanotechnology.

In recent years, consumers have increased their awareness and demand for functional foods with impactful therapeutic benefits. As such, food manufacturers have enhanced their focus on adopting processing methodologies that would be of minimal deleterious effect on bioactive components in food products. Several nonthermal approaches have emerged and are greatly improved to fix these concerns.

On another note, the nanoscale manipulation of food ingredients at the industry level has also been observed to be of significant benefit in improving their biological functionality and ensuring their biodelivery to bodily sites of action subsequent to ingestion. Accordingly, the need arises to integrate the nanotechnological concepts with emerging food processing technologies for the development of food products with optimal dietetic and health benefits.

Emerging innovative techniques within the food industry including nanosensors, nano-food packaging, nano-bioremediation, and efficient processing of bioactive components in foods when executed at nanoscales are discussed in this book in detail.

This book is a comprehensive evaluation of the latest food-nanotechnological developments with an emphasis on various applications in different food industries. It describes the most recent techniques in depth and includes a variety of important sources that make this an ideal resource for researchers in this captivating world of food nanotechnology.

CHAPTER 1

General Overview of Nanoscience in Food Technology: Basics and Research

PAPIHA GAWANDE[1], POOJA KESARKAR[1], and YOGESH GAT[2*]

[1]*Department of Food Engineering and Technology, Institute of Chemical Technology, Marathwada Campus, Jalna, India*

[2]*Department of Food Engineering and Technology, Institute of Chemical Technology, Mumbai, Maharashtra, India*

*Corresponding author

ABSTRACT

Nanotechnology is emerging as an intriguing subject in food technology, with several opportunities to improve food quality, safety, and sustainability. This chapter presents a general overview of the application of nanoscience in the food sector, concentrating on its fundamentals and ongoing research. Nanotechnology is the manipulation and control of matter at the nanoscale, which typically ranges between 1 and 100 nm. Nanotechnology, when applied to food technology, provides several key principles and approaches that drive innovation and advancements in various parts of the food industry. The incorporation of nanotechnology into food processing, packaging, and delivery systems has transformed the area, allowing for the development of novel solutions. Nanoparticles, such as nanoemulsions and nanocapsules, have demonstrated tremendous promise in terms of nutrient delivery, flavor encapsulation, and antibacterial characteristics. Furthermore, nanosensors and biosensors have enabled the rapid and sensitive detection of pollutants, pathogens, and quality indicators, hence improving food safety and traceability.

Nanotechnological Advancements in the Food Industry.
Ishrat Majid, Bababode Adesegun Kehinde, and Sheikh Shafat Hussain (Eds.)

DOI: 10.1201/9781774919750-1

1.1 INTRODUCTION

Many demanding sections of the agriculture and food industries have considerable potential for applying nanotechnology, which is creating new methods for food production, fertilizer distribution, package manufacturing, plant cropping, animal feeding, and animal culture, among others. Nanotechnology has transformed several sectors, and it is currently making significant advances in the food business. With the world's population continually increasing, the problem of feeding everyone has grown more difficult. By considerably expanding the shelf life of food components and enabling global distribution, nanotechnology has emerged as a major changer. While it is not a perfect solution, it has significantly decreased food waste caused by microbial infection. The food industry has discovered a feasible way to handle the critical issue of preserving food and ensuring it reaches a broader population by utilizing nanotechnology, therefore addressing the requirement for sustainable food production. Foods that have been presented to consumers using nanotechnological tools and techniques, and/or contain a mix of nanoparticles in an acceptable range are referred to as "nano foods." At the nanometric scale (10–100 nm), the structure and properties of matter can be altered, opening up novel and exciting study fields where nanotechnology and biology are integrated. By altering the physicochemical characteristics of the food's nanoscale ingredients, nanotechnology enables the improvement of food product quality. Advanced science known as nanotechnology is already having an impact on a variety of other industries, such as agriculture, food, cosmetics, and medicine. Additionally, the study of how the structure, texture, and quality of foods can be altered by the physicochemical properties of nanometric materials is being done in the field of food nanotechnology. Food antimicrobials are being developed, such as new food packaging materials with improved mechanical and antimicrobial properties, barriers, observing the condition of food during storage and transport nanosensors for traceability, and the encapsulation of food components or additives. Additionally, food nanotechnology encompasses developing food antimicrobials; increasing the bioavailability and absorption of nutraceuticals and health supplements; and improving the texture, flavor, color, and consistency of foods (Chaudhry et al., 2008). Due to the numerous benefits that these technologies can offer the encapsulated material, the spectrum of applications for microencapsulation and nanoencapsulation in the food business has been expanding. These include providing buffering against temperature, high pH, and ionic strength variations, masking off odors or tastes, and by shielding the encapsulated material from a variety of chemical, environmental, and enzymatic changes, it can be made more stable.

1.1.1 RESEARCH TRENDS OF NANOSCIENCE IN FOOD TECHNOLOGY

Food nanotechnology and its application in human society are currently in dire need. Science is experiencing enormous, yet far-reaching, scientific obstacles. Nanomaterials are used in food packaging to extend shelf life, cut down on food waste, and maintain food quality. The bioavailability and nutritional value of bioactive substances are improved when they are encapsulated in nanoparticles. Food ingredients can be altered via nanotechnology to enhance their texture and sensory qualities. The efficiency of additives is ensured, and the necessary amount is decreased, by their controlled release using nanocapsules and nanocarriers. Real-time data on food quality and traceability is provided via intelligent packaging with nanotechnology elements. Nanotechnology further provides tailored pesticide and fertilizer delivery, supporting sustainable food production. However, the use of nanomaterials in food presents safety and regulatory problems, calling for careful study and the creation of relevant legislation. In general, food nanotechnology holds the promise of revolutionizing the food sector by assuring safer and more nutrient-dense products while minimizing waste and adverse environmental effects.

Gupta and Tomer (2017) dealt clearly and persuasively with recent advancements in nanotechnology in a thorough assessment. The trademarks of this thoroughly researched review project are scientific fortitude, profundity, and tolerance. Nanotechnology has a wide range of applications in several fields of scientific study. By manipulating matter at the nanoscale and making use of novel phenomena and properties (physical, chemical, and biological) at that futuristic length scale, nanotechnology creates useful materials, advanced materials, and systems. All aspects of dental practice are being transformed by new research and technology, which has altered conventional methods for diagnosis, risk assessment, preventive, and many other procedures.

This century's most recent technology, nanotechnology, is having a huge impact on a wide range of industries, including food processing, pharmaceuticals, agriculture, and many more. On the way to scientific renewal in this century, technological validation and scientific perspective are absolutely necessary. There is still an intact umbilical cord connecting the enormous problems, the accomplishments of sustainability research, and the science of food nanotechnology today. Nanotechnology uses in the food industry are entirely distinct from their varied applications in the medicinal sector.

1.1.2 CURRENT AND PROSPECTIVE APPLICATIONS OF NANOTECHNOLOGY IN THE FOOD AND AGRICULTURE SECTOR

1.1.2.1 NANOBIOSENSORS AND NANOSENSORS

In the food industry, nanosensors and nanobiosensors are used to monitor food processing, food packaging, food quality evaluation, food storage, contamination in food, and food safety. In order to assess data on a very small scale for real-time biologically based analytical applications, nanobiosensors are built. The major inferences in the area of agriculture are soil quality and fertility, physical monitoring of temperature and humidity, sensing microbiological microenvironment of the soil, response sensors for irrigation and safety in agronomy, and indicator for seed viability and shelf-life, detection of residual pesticides, precision agriculture, and toxins and fertilizers, plant pathological monitoring, and so on. Probes are antibodies, receptors, lectins, enzymes, tissue, nucleic acids, microorganisms, organelles, molecular imprints, and so forth. The transducer functions as a sensor and is made of nanomaterials such as carbon nanotubes, graphene oxide, quantum dots (QDs), iron oxide, and metallic gold nanoparticles.

1.1.2.2 WORKING PRINCIPLE OF NANOSENSORS

Nanosensors have the ability to measure as low as a single molecule. An analyte, a transducer (a component that converts one form of energy into another), a sensor, and a detector make up nanosensors. Typically, nanosensors work by keeping an eye on electrical changes in the sensor materials (Figure. 1.1). An accurate and effective reaction occurs when the analyte diffuses from the solution to the surface of the sensor.

This change in the surface's physicochemical properties results in the transducer's optical or electrical characteristics changing. This alteration is then transformed into an electrical signal that is detected. Table 1.1 includes various kinds of nanosensors used in food industries.

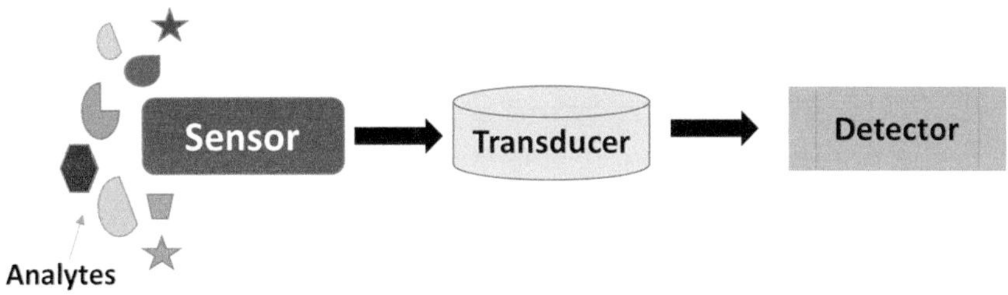

FIGURE 1.1 Working mechanism of nanosensors.

1.1.2.3 ELECTROCHEMICAL NANOSENSORS

The electrochemical nanosensors depend on chemical reactions between the target analyte, biological element, nanofabricated chemicals, and biomolecules, to make or consume ions or electrons that are then detected as current, impedance, or voltage. The amount of analyte contained in a sample was quantitatively associated with the electrochemical signal produced. Electrochemical nanosensors can be divided into amperometry, voltammetry, and potentiometry devices based on how they function. An electrochemical sensor typc called an amperometric sensor continually measures the current produced by the redox reaction of an electroactive species.

Voltammetry is the process of measuring current after a controlled change in potential has been made. The electrochemical reaction rates and redox potential of the electrochemical reaction with the analyte are best obtained using cyclic voltammetry. Potentiometric sensors gauge the potential difference between the reference electrode and the working electrode. The output signal is produced as a result of the equilibrium ion buildup at ion-selective electrodes and ion-sensitive field-effect transistors.

1.1.2.4 OPTICAL NANOSENSORS

The analytical performance of the optical nanosensors can be achieved using a spectrofluorometer or a fluorescence microscope with quantified fluorescence detection. Known concentrations of the analyte are then

TABLE 1.1 Various Kinds of Nanosensors Used in Food Industries.

Nanosensors	Types of Nanosenors	Properties/Functioning	Application
Electrochemical nanosensors	Amperometry, Voltammetry, Potentiometry	• Helps in faster and more generous transfer of electrons from the surface to the detector and a receptor for analyte interaction.	Detection of pathogens, toxins and even viruses.
Optical nanosensors	Surface plasmon resonance (SPR), Surface-enhanced Raman scattering (SERS), Quantum dots(1–10 nm)	• Agglomeration of nanoparticles upon bacteria-GNP interaction.	Estimation of enteropathogen Salmonella typhimurium.
		• Ability to have excellent and very stable photoluminescent properties.	Diagnose pathogens and their related toxins in food and water
		• Fluorescence-based sensors have the ability to cancel out background nonspecific signals, increasing the system's sensitivity and assisting in the detection of tiny compounds such as pesticides	Able to detect as less as 1 nM of aflatoxin B1, a secondary fungal metabolite of Aspergillus flavus. To detect a multitude of contaminants ranging from bacteria like Pseudomonas to even small antibiotic molecules present in food samples.
Nanobarcode technology	RFID technology	• Interpret the encoded data or information in the form of a map.	Barcodes for rapid and robust detection of Salmonella enteric Serovar Enteritidis in the food sample.
E-nose and E-tongue	Metal oxide sensors, Conductive polymer sensors, Quartz crystal microbalance sensors, Optical sensors, Surface acoustic wave sensors, Gas-sensitive field effect transistors	• A chemical sensor array composite that alters conductance when in contact with a volatile analyte and sends a detection signal to the acquisition system.	Monitoring of raw and manufactured product, cooking and fermentation process, product packaging, storage, and quality assessment for long-term storage Evaluation of aroma transfer from food plastics bags Detection of Alicyclobacillus acidoterrestris spawned spoilage in apple juice, etc.

introduced to the fluid in which the nanosensors are suspended. Due to the small size of the nanosensors, the aqueous phase does not scatter light so appears transparent when nanosensors are suspended in it.

In general, a wide range of optical techniques based on surface plasmon resonance (SPR), fluorescence spectroscopy, interferometry, and spectroscopy have been used in biosensors. Luminescence contains dual components: fluorescence and phosphorescence. The effects of fluorescence, an intrinsic feature of gold, silver, and QD nanoparticles, depend significantly on the size and morphology of the nanoparticle. In the form of binary compounds such as GaAs, CdSe, CdTe, InAs, and SiC as well as ternary compounds like QDs, InGaN, InGaP, and InGaAs, are nanoclusters of a few hundred to a few thousand atoms.

SPR is a different optical sensing type used for extremely sensitive and quick detection. SPR biosensors have gained popularity for studying a variety of analyte molecules because of their distinctive features, including real-time monitoring, high specificity, sensitivity, and no requirement for labeling.

1.1.2.5 NANO BARCODE TECHNOLOGY

By decoding data or information and translating it into a map-like representation, barcode technology adheres to the notion of symbology. A collection of small DNA sequences from various species is used in the process, and the data is analyzed by building a phylogenetic tree using the distance-based neighbor-joining method.

1.1.2.6 E-NOSE AND E-TONGUE

To ensure optimum consumer satisfaction, the e-tongue and e-nose are increasingly being used as replacements for the consumer panel and human sensory expert panel formed for quality control and quality assessment during the manufacturing processes. The sample headspace system, an electronic data acquisition control system, pattern recognition software, and a sensor array are the four parts that make up the e-nose. In the manufacturing and processing of food, drinks, agricultural products, pharmaceuticals, and personal care products, the flavor and aroma of volatile and nonvolatile ingredients retain important information about the quality and quantity of the substance (Baldwin et al., 2011). Recently, the e-nose and e-tongue have been utilized to analyze the quality of beef fillets and the evaluation of scent transfer from food plastic bags (Mohareb et al., 2016), detection of mixed edible oil (Men et al., 2014), detection of Alicyclobacillus acidoterrestris spawned spoilage in apple juice and identification of adulterated milk.

1.2 APPLICATION OF NANOSENSORS AND NANOBIOSENSORS

Recent years have seen the emergence of nanoparticle-based sensors as new tools for improving agricultural methods. Table 1.2 includes the important potentials of nanosensors in the agricultural sector. The word "nanosensors" refers to a particular class of designed nanoparticle-based sensors that can continuously monitor the environment and the health of intact plants. These sensors built of nanoparticles enter and move through the intricate biofluids of plants, where biomolecules such as metabolites, proteins, carbohydrates, and lipids adsorb and form a coating known as the "bio-corona" on the nanoparticle surfaces. For achieving the development and widespread use of nanoparticle-based sensors in agriculture, it will be crucial to comprehend these interactions between nanoparticles and biomolecules that control nanosensor activity in plants (Voke et al., 2021).

TABLE 1.2 Important Potentials of Nanosenors in Agriculture Sector.

Potential	Description
Detection of gases in atmosphere	Metal oxide-based nanostructures can be integrated with existing gases and technology.
	Possible to identify the gas composition for usage in environmental, industrial, and domestic settings.
	Able to detect minute amounts of certain gases in the environment.
Track water, soil, and environmental changes	Creates a powerful field-effect transistor by integrating nanowire arrays on a small semiconductor chip.
Detect contamination of food	Detect infections in order to avoid food contamination.
	Important tool for tracking dynamic biomolecular changes and biological functions, containing crucial data that aids in our understanding of basic biological mechanisms.

Due to each material's distinct and favorable features as well as their synergism for use in sensing applications, graphene-based hybrid materials have recently attracted a lot of attention. In addition to offering advantages over traditional analytical techniques, including other graphene-based nanomaterials, graphene-based hybrid sensors provide a selective sensing platform. It is anticipated that graphene and NP hybrid films would be able to offer a number of multifunctional features in addition to pushing the boundaries of analytical sensitivity and selectivity that are currently possible. We may anticipate that graphene with NP hybrid structures would be helpful in electronic, optical, electrochemical sensing methods for the identification of chemical and biological compounds, as well as complete living cells in the field of chemical and biological sensing (stem cell differentiation) (Sundramoorthy et al., 2018).

The efficient immobilization of substances on surfaces is governed by graphene oxide and functionalized carbon nanotubes, which opens the way for the development of enzyme-based biosensors in addition to numerous uses for nanosensors and nanobiosensors based on metal nanoparticles. Close confinement facilitates effective electron flow between enzyme and electrode and streamlines the immobilization stages, enhancing repeatability and operability. It also has the advantage of controlling protein orientation by eliminating randomly bounded proteins to increase sensitivity. Carbon nanotubes that have been amino-functionalized are used to create an extremely delicate organophosphorus pesticide biosensor electrode by effectively immobilizing AChE on the surface of the glassy carbon electrode. With a 0.08 nM limit of detection, these electrodes have been effectively used to detect paraoxon and other pesticides from vegetable samples (Yu et al., 2015).

Using a hyperbranched polyethyleneimine-based silver nanoparticle-based nanosensor, the carcinogenic nitrile contaminant was discovered. Fluorescence quenching has been linked to nitrite concentrations below the 100 nM limit of detection, which causes the reaction of nitrite with H_2O_2 to produce peroxynitrous acid in acidic circumstances and stimulate the aggregation of silver nanoparticles (Chen et al., 2016).

1.2.1 NANOSENSORS IN FOOD PACKAGING

One of the most creative ideas is active packaging, which is also the major force behind the use of nanomaterials in food packaging (Figure 1.2).

This method is created on the idea of adding particular components of antioxidants and additives with antimicrobial, and also fungicide properties into the packaging systems, in an effort to extend the life while maintaining or enhancing the quality and condition of the packaged food.

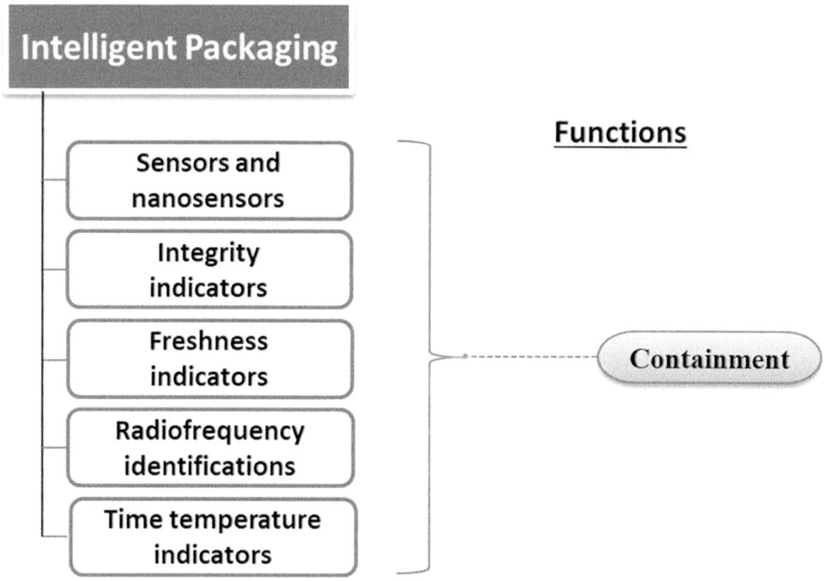

FIGURE 1.2 Application of nanoscience in intelligent packaging.

By ensuring the quality indicators, the IP are technologies that keep an eye on and deliver data about the condition of packaging, food storage, or the internal and exterior environment. They are regarded as the next generation of packaging systems. These packaging systems have the intelligence to carry out intelligent tasks like acquisition, detection, recording, communication, monitoring, and the application of scientific reasoning to aid in decision-making, increase lifespan, and forewarn possible issues. Without generally affecting the food, this sort of packaging examines the system, processes the data, and shows it. Most control devices allow consumers to check the quality of food through a color response that matches or relates to the quality of food at a specific temperature. They are easy to use and may be included in the package. They may be based on physical systems like diffusion, biological systems like enzyme reactions, or chemical reactions like polymerization processes, depending on the mechanism at work. They might also need some initial activation, such as by ultraviolet radiation, or some combination of these systems.

The most typical are the leakage indicators, which are used to find perforations and other problems with sealed containers. The two most often used indicators are gas and oxygen (O_2) and carbon dioxide (CO_2). Time indicators, which reveal how long a product has been opened, are the more straightforward integrity indicators. When food quality deteriorates during storage, transportation, sale, and consumption, freshness indicators are supposed to send out a signal that the product is no longer suitable. The signal is based on a change in the indication that happens as a result of various compositions in the package's atmosphere brought on by a product's chemical and microbiological alterations.

Radio-frequency identification (RFID) is a wireless data-gathering technology that communicates among a transmitter and/or receiver using electromagnetic waves (EW). These RFID tags are made up of an antenna and a microchip memory storage unit that are implanted in a substance called substrate. RFID tags can be used for intelligent food packaging, allowing for antitheft protection, control of integrity, authenticity, traceability, anticounterfeiting, and quality.

1.2.2 FOOD PRESERVATION AND PROCESSING

Food processing is the process of transforming raw components into edible and commercially viable food products. It includes operations such as detoxification, microbiological protection, preservation, and quality enhancement, allowing for improved distribution and marketability. Processed foods are well-suited for long-distance transportation and frequently have higher resilience to deterioration than fresh meals. Food processing converts basic ingredients into safe, long-lasting, and handy products, maintaining efficient supply chains and providing customers with more nutrient alternatives.

Nanotechnology has had a substantial influence on a wide range of consumer goods, with nanoparticles being used in a variety of items. Many people are unaware of the existence of these nanoparticles, which have revolutionary properties such as easy-to-clean and scratch resistance. Samples of automobile bumper are lighter in weight, clothing is a lot of stain repellent, sunscreen has improved radiation resistance, as have cell phone screens, synthetic bones, and drink containers made of glass result in an extended shelf-life, and balls for varied sports are created a lot of sturdy. Exploitation of nanotechnology, within the mid-term embedded "wearable electronics" allows for the "smart" transformation of fashionable textiles. These innovative products hold great promise, particularly in cosmetics, and possess significant potential for diverse applications across various industries. Technology is expected to play a significant role in driving both business and technology in the twenty-first century. It promises higher-performance materials, intelligent systems, and new production techniques that will have a major impact on all facets of society.

1.2.3 NANO CAPSULES

Encapsulating emulsions, simple solutions, biopolymers, and colloids into meals maintains the relevant properties since delivery systems are crucial in the process industry. A liquid transporter of healthful components that are water and fats insoluble is a nano-sized self-assembled structured lipid known as a "nano-drop." They are used to stop cholesterol from entering the bloodstream after leaving the digestive tract.

Nanotechnology offers economical and long-term solutions to challenging technical and scientific issues in food and bioprocessing. It facilitates the production of safe and high-quality food using innovative and environmentally friendly methods. Food quality watching victimization biosensors and Bacterium identification; active, sensible, and intelligent food packaging systems; nanoencapsulation of bioactive food complexes are a few examples of growing applications of nanotechnology for the food trade. Nanotechnology is applied within the processing, production, packaging, and safety of food. Anti-microbial agents placed directly on the coated film surface of food packaging using a nanocomposite coating process could enhance packaging. The flexibility of nano-composites allows for the modification of the gas permeability of fillers based on individual product needs, either increasing or lowering it. They will conjointly lower the gas transmission rate and enhance mechanical and heat-resistant qualities. Analysis is being done to use engineering science for the detection of chemical and biological compounds aimed at food sensing.

While adulteration of food products is normally forbidden, new food items using nanotechnology are rapidly reaching the market at an average of three to four each week. Canola Active Oil, a canola vegetable oil infused with "nano drops" to aid in the transport of minerals, vitamins, and phytochemicals via the digestive system, and nanoceuticals slim shake chocolate, a chocolate diet drink, are two examples. These goods make use of nanotechnology to improve their functionality. The packaging of these nanotechnology-based food products has a number of advantages. For starters, it offers effective food packaging, which ensures product safety and preservation. Second, it acts as a barrier, keeping gases, moisture, and pollutants from permeating. Furthermore, the packaging provides antimicrobial protection, limiting microbe development and increasing product shelf life. It is also meant to be biodegradable, reducing environmental effects. Finally, the packaging has smart capabilities that might allow for real-time monitoring of food quality and freshness. These products use nanotechnology in food packaging to improve food safety, lengthen shelf life, and improve overall customer experience.

Active and intelligent packaging has been developed with the development of nanotechnology and the wide range of qualities that NPs possess. Preservatives from the container have been released into the product, extending its shelf life and lowering the concentration of chemical preservative systems inside the product, enhancing its security. Consumer demand has driven the development of packaging materials such as coatings and films that are entirely composed of biodegradable chemicals known as biopolymers. This trend toward sustainable packaging alternatives is consistent with rising environmental awareness and demand for eco-friendly solutions. Biodegradable packaging has positive effects on the environment and environmentally friendly trash management. The carrageenan, biopolymers cellulose, agar, chitosan, and starch have all received the most research attention. However, much research came to the conclusion that employing biopolymers in packaging manufacture has disadvantages, particularly in terms of permeability and strength, sometimes known as barrier and mechanical qualities. Nanocomposites are systems that are demonstrated to increase these characteristics.

A nanocomposite is a heterogeneous arrangement made up of two or more ingredients having distinct chemical and physical properties, at least one of which has a nanoscale size. Biopolymers serve as the continuous phase in packaging-based nanocomposites, while nanofillers serve as the discontinuous phase. Biopolymers offer unique qualities, which make them perfect resources for making active antimicrobial films, such as "bacterial" nanocellulose, when they are manufactured at the nanoscale. This is a natural nontoxic biopolymer with high porosity, surface area, and crystallinity, and resilience that is produced from certain bacteria with a distinctive nanostructured morphology. Numerous compounds, including antibacterial drugs, can be retained by the hydroxyl groups of bacterial nanocellulose within its pores. Chitin, which comes from shellfish waste and is the second most common polysaccharide resource in nature, varies from cellulose where at the C-2 position it has an amino group rather than a hydroxyl group. Chitosan, which is its largely deacetylated state, can be functionalized to create derivatives with cationic or other units. Chitosan and its derivatives have antibacterial action that depends on their degree of acetylation, cationic charge density, position, molecular weight, hydrophobicity, and any other functional groups they may contain, among other structural characteristics. Due to the differences in the composition of their cell envelopes, gram-negative and gram-positive bacteria exhibit different modes of action. The cationic charges in the chitosan backbone provide the basis for the antibacterial activity in both situations. NPs, nanorods, nanotubes, and nanofibrils have all been used as nanofillers in these composites.

Despite its infancy, the use of nanotechnology in the preservation of food and cosmetics, particularly in food packaging, provides considerable benefits, as evidenced by various studies. Its application in the food business is still in its early stages. Both sectors have used a wide range of nanostructured materials, including metallic nanoparticles and their nanocomposites, as well as new nano-based delivery systems for natural preservatives, revolutionizing many areas of scientific knowledge in those professions to date. For instance, silver nanoparticles are anticipated to be crucial in the design of polymeric materials. They are extensively employed in active food packaging, with bioplastics expected to replace nonbiodegradable polymers. This transition is anticipated to be complemented by the integration of nanocomposites, aiming to extend shelf life and improve the overall quality of food products.

Nanoclays, which are extracted minerals, metallic nanoparticles from bio-fermentation, and bioplastics made from renewable biomass sources might all be used to make sustainable smart food. Chemical preservatives' negative effects are eliminated by the on-demand release of preservatives from active packaging, and the retention of natural antibacterial agents encourages the use of more natural products. Fruit juices and other liquid meals can be heated to produce beverages of the finest quality by substituting nanofluids for water, and nanomilling is a practical technique to increase the bioavailability and preservation action of many natural components. Nanotechnology could also be used to address one key problem that has an influence on the environment, the economy, and public health: food waste. While enhancing product quality, the risk of contamination of products will be reduced, and packaging waste will not affect the environment.

1.3 QUALITY CONTROL AND QUALITY ASSURANCE

NMs are thought to be one of the newest items that benefit people. In a variety of fields, it is anticipated that the use of NMs would significantly rise. Long-term contamination of natural resources is encouraged by the use of NMs in the fields of medicine propulsion systems, food industries, cosmetology, agriculture, and personal healthcare systems. The possibility exists that, eventually, NM residue will enter the human food chain. The whole economy might easily accelerate at the same time if NMs were successfully used for a variety of applications. It is important to evaluate both the advantages and disadvantages of applying these NTs to each area. To ensure that any nanomaterial is safe for use in the food business and is not harmful to people or the environment, a thorough investigation of its toxicity is required before it can be approved for operational use in this sector.

1.3.1 ADVANCEMENTS IN NANOMATERIAL-BASED SENSORS: REVOLUTIONIZING FOOD QUALITY ASSESSMENT

Foodborne illness occurrences are a constant challenge for the food engineering industry. Foodborne disease is defined as an ailment that is often contagious or fatal in nature and is brought on by agents that enter the body through food digestion. Protozoa, bacteria, viruses, bacterial toxins or fungi, insecticides, and metal ions are a few examples of the contributing creatures. *Brucella, Bacillus cereus, Escherichia coli, Campylobacter jejuni, Clostridium perfringens, L. monocytogenes, Shigella* spp., *S. aureus,*

Salmonella typhi, S. paratyphi, V. parahaemolyticus, and *Vibrio cholera* are a few microorganisms linked to serious foodborne illnesses. It is crucial to develop quick, low-cost ways to detect the presence of harmful compounds or the presence of pathogen, contamination in food in order to manage the prevalence of foodborne disease. Both the PCR-constructed methods and the immunoassay-based processes are common methods that are consistently used. There is no question that these approaches are scientific, but they require expensive equipment and chemical components, specialized personnel, extensive sample preparation, and drawn-out procedures, which impede preventive patient management. The creation of novel methods that are simple to use, speedy, affordable, and sensitive for site-specific recognition as well as stable and mobile sensing kits is urgently needed. In the previous period, NT was able to cover certain strategies in this area (Hossain et al., 2021).

1.3.2 EXPLORING CHARACTERIZATION TECHNIQUES FOR NANOMATERIALS IN FOOD MATRICES: UNLOCKING INSIGHTS INTO FOOD SAFETY AND QUALITY

The development of nanotechnology made it possible to apply various NMs in the fields of healthcare, ophthalmology, biotechnology, spectroscopy, energies, and electronics in addition to the food sector. Size, shape, zeta potential, and polydispersity indices are among the traditional methods used to characterize nanomaterials. The fundamental understanding of the size, shape, charge, dimension, surface area, structure, and porosity of the nanocomposites that can be known by chemical classification is aided by these characterization technologies. However, improvements in characterization technique and resolution ranges are necessary to comprehend the physicochemical characteristics of nanocomposites (Yu et al., 2021).

For estimating the dietary matrix, many spectroscopic and electron microscopic techniques are helpful. Biomaterials are typically included as polymers that improve the material's capacity for resolution. X-ray fluorescence, X-ray absorbance, nuclear magnetic resonance, FTIR, and UV-visible spectroscopy are among the main spectroscopic methods employed in the analysis of biomaterials and NMs. To understand the morphology, size, and interaction of the NMs, however, sophisticated microscopic characterization techniques are also employed. A thorough description of NMs can be provided utilizing skimming investigation nuclear power microscopy, microscopy, transmission and scanning electron microscopy, and confocal laser scanning microscopy, and field radiation skimming electron microscopy.

A compound's elemental map and electronic structure can be determined using X-ray absorbance spectroscopy. Understanding the molecular structure, quality assurance, and purity of the samples are accomplished through the use of nuclear magnetic resonance spectroscopy. Understanding the sample's chemical and structure bonds that were formed during its creation is made possible by FTIR spectroscopy. In order to quantify the concentration and diffusion of certain Nanomaterials in the form of resolution, UV-visible spectroscopy is used. Understanding biological interaction and imaging is made possible by microscopic methods like electron microscopy and atomic force microscopy. All of these techniques aid in the assessment of various food matrices to evaluate different biomolecule potential for improving our knowledge of the biomolecules consumed by humans (Sekhon, 2010).

Enhancing several elements of food quality is the main goal of using metal nanoparticles (MNPs) in agriculture and the food business. These applications are mostly focused on enhancing the flavor, color, and texture of food. In order to promote nutrition absorption and improve the body's ability to use critical nutrients, MNPs are also used. Additionally, they are essential for the targeted distribution of bioactive substances and nutrients, ensuring that they reach certain tissues or cells in the body. MNPs also aid in the creation of active packaging technologies that stabilize food items, enhancing product quality and extending shelf life. Finally, MNPs act as sensors to track and uphold food safety and quality standards, protecting the health of customers (Couto and Almeida, 2022). Table 1.3 describes the application of metal nanoparticles in the food sector.

1.3.3 NANOTECH IN NEUTRACEUTICALS (NANOLIPOSOMES)

As an organizing model for nourishing well-being and alleviating health distress, nutraceuticals need to be given a significant place in the food and drug system. The success of these goods was attributed to a number of factors, including the use of new medications and their partial success in treating severe illnesses that could push society toward other effective alternatives to older medications with negative side effects. A stronger indication of nutritional research and pharmacological or epidemiological revisions taken into account for the welfare and efficacy of phytotherapy finished through widespread among those who look for dietary improvements to prevent illness. Nutraceuticals are frequently divided into three broad categories, such as dietary advancements, herbals, and nutrients. They continue to be used on a daily basis by people, for example, as an alternative to modern medicine, supporting a full and fulfilling existence, as seen in. They must be decided to offer advantages comparable to those of a regular immune booster and antioxidant, as well as fewer side effects as compared to drugs, improved bioavailability, and a long half-life. The significance of the document is partially due to a variety of vitamin reasons, specifically liposoluble vitamins that have been nano-formulated. Many researchers continued to make significant progress in relation to the useful bioavailability and stability of vitamins, hence validating health benefits among consumers.

The formulation, production, and procedures of nutraceuticals and functional foods have made use of some nanoscale events. To increase product functionality and distribution effectiveness, new ideas based on nanotechnology are being investigated. The use of nanotechnology in nutraceuticals is growing swiftly because it can improve the solubility, stability, and bioavailability of active substances. Their distinctive characteristics are mostly caused by their diminutive size and high surface-to-volume ratio. Any unfavorable effects that nanoparticles may have on human health are also due to their physicochemical properties. It is vital to assess the effectiveness and safety of novel commercially viable nutraceutical formulations. Time and money are saved with this tactic.

Hydrophobic interactions lead to the spontaneous self-assembly of a wide variety of thermodynamically stable, liquid crystalline phases, lyotropic with characteristic length scales in the nm in amphiphilic molecules such as surfactants, copolymers, and lipids, that have nonpolar and polar properties. These include hexagonal (tubular) structures, micelles, cubosomes, and lamellar structures which exist in a liquid state but have a high degree of molecular orientation order. Micelles are submicron spherical particles that spontaneously form when surfactants dissolve in water at concentrations above a critical level, or "critical micelle concentration." They range in size from 5 to 100 nm (CMC). Surfactant self-assembly is thermodynamically driven, with the goal of minimizing contact between the hydrophobic tail group and water while maximizing interactions between the hydrophilic head groups and water. This permits micelles to maintain their shape for long periods of time, despite changes in pH, temperature, and salt content. Micelles are extremely effective for encapsulating nonpolar substances such as lipids, flavorings, antimicrobials, antioxidants, and vitamins. Because of this unique property, they can successfully preserve and distribute these chemicals in aquatic settings. Micelles serve an important function in solubilizing compounds that are insoluble or sparingly soluble in water. Microemulsions, which are micelles containing solubilized components, improve this capability even further. While micelles have typically been utilized as pharmaceutical component carriers, their potential as functional food ingredient delivery systems has received considerable interest in recent years. This new application enables the effective integration and regulated release of functional food components such as bioactive chemicals and minerals, improving their bioavailability and possible health benefits. Several effective microemulsion applications include the encapsulation of lycopene, limonene, omega-3 fatty acids, and lutein utilizing food-grade emulsifiers. In other circumstances, however, the use of ethanol as a co-surfactant was necessary. Patent applications have shown how microemulsions may be used to encapsulate alpha-tocopherol, successfully minimizing lipid oxidation in fish oil. Furthermore, microemulsions have been used to infuse essential oils into flavored carbonated beverages, providing a fresh way to improve their scent and taste. These developments emphasize the versatility of microemulsions as a bioactive component delivery technology, offering increased stability, controlled release, and enhanced functioning in a variety of food and beverage applications.

1.3.3.1 LIPOSOMES

Liposomes or lipid vesicles are made from polar lipids, chiefly phospholipids derived from soy and egg, which are common in nature. Liposomes, like micelles, can include a variety of functional components. Unlike micelles, they can encapsulate both water and lipid-soluble compounds. Liposomes are spherical polymolecular aggregates with a bilayer shell structure. Lipid vesicles can be unilamellar, with a single bilayer shell, or multilamellar, with numerous bilayer shells, depending on the production mechanism (Honda et al., 2013). Liposomes are generally 20 nm to a few hundred micrometers in size, with an inner composition similar to the aqueous solution used to generate the vesicles. According to the charge of the polar lipids used in the creation of liposomes, charged and water-soluble ionic species could be contained within them. Thus, the pH and ionic strength of the liposomal core can change in comparison to the continuous phase in which the liposomes are later disseminated. Liposomes encapsulate proteins while generating a microenvironment that ensures their functional stability regardless of external circumstances. The inside of the lipid bilayer, on the other hand, has characteristics similar to those of an organic solvent. As a result, lipid molecules can be solubilized and contained inside the bilayer. Phosvitin that has been liposomally-entrapped was employed to prevent lipid oxidation in a range of dairy products and ground pork. Liposome-encapsulated vitamin C preserved 50% of its action after 50 days of refrigerated storage, but free ascorbic acid lost all activity after 19 days.

1.3.3.2 ADVANTAGES OF NANOMATERIALS IN NUTRACEUTICALS

Due to their purported safety, potential nutritional value, and medicinal effects, nutraceuticals have attracted a lot of interest. Herbal nutraceuticals are useful for maintaining health, as well as for acute or chronic illnesses that are triggered by dietary deficiencies. They also aid with aging, quality of life, and health. Public education, production, processing, a renewable source, environmental friendliness, and local accessibility are growing trends in nutraceuticals. Enzyme, analyte, and related dye interactions can be managed by nanosensors. Packaging that incorporates nanosensors designed to attract and detect biochemical and microbiological change. Nutraceuticals and bioactive substances that are effectively solubilized, stabilized, encapsulated, and delivered via nanotechnology have benefits including improved therapeutic index, increased absorption in low dosages, and decreased frequency of dose administration.

1.3.3.3 NANOBIOCATALYST (NBC)

NBCs combine an enzyme, a biological component for biocatalysis, with a nanomaterial carrier that possesses special electrical, optical, magnetic, and stimulus-responsive capabilities. There have been many reported technologies for producing specific and processable enzymes, including recombinant DNA technology to reduce the cost of commercial enzymes, directed evolution to pursue specific traits and expand the repertoire of substrates, and bioprocesses to produce recyclable and long-lasting enzymes for industrial applications.

NBCs can be created by immobilizing enzymes on or encasing them in a variety of nanomaterials, such as metals, silicas, polymers, and carbons. These methods include covalent binding, cross-linking of enzymes, electrostatic interactions, hydrophobic interactions, van der Waals forces, or hydrogen bonds as well as physical entrapment or encapsulation. NBCs can be created using a variety of techniques, based on the physical and chemical characteristics of the nanocarriers and enzymes, as well as the uses and processes for the finished products.

1.3.3.4 NANOENCAPSULATION

The public must be made aware of the advantages of adopting nanotechnology for other objectives than present technologies if the food and agriculture industries are to fully use its potential. These methods are utilized to effectively transfer compounds into the cells of plants and insects (Scrinis and Lyons, 2007). Nanocapsules are capable of carrying out effective chemical penetration through tissues and cuticles, resulting in a gentle and continuous release of the active substances. Specific nanoencapsulated pesticides have been developed in comparison to conventional pesticides to only kill the desired insects, hence lowering the effective dose. Nanoencapsulated pesticides are adsorbed to the surface of the plant, allowing for their sustained release, which takes longer, whereas traditional pesticides are washed away in the rain. Similarly, DNA and other compounds can be delivered through nanoencapsulation into plant tissues to shield them from insect pests (Torney et al., 2007). The mechanisms for releasing encapsulated components are diffusion, dissolution, osmotic pressure and biodegradation with a specific pH. Eutrophication is a long-standing issue that can be solved by using nanoencapsulation techniques to improve the effectiveness of fertilizers. Fertilizers can be nanoencapsulated in three different methods. The nutrients might be given as nanoscale particles or emulsions, covered with thin polymer films, or enclosed inside nanoporous materials (Ding and Shah, 2009).

1.4 CHALLENGES OF NANOSCIENCE

Nanoparticles' tiny particle size and large surface area, which allow for facile dispersion, may allow them to circumvent anatomical barriers, reach more distant parts of the body, and demonstrate potential toxicity, are the key factors contributing to the risk of employing this technology. Farmers' health may be in danger due to the handling of nanoparticle pesticides and fertilizers, which are quickly absorbed into the soil, water, or environment. Additionally, there is a worldwide discussion about the merits and disadvantages of employing nanosilver in healthcare goods as an antibacterial agent. In the food industry, particle nanomaterial migration from food contact materials into food may pose potential dangers to human health. This is dependent on migration, consumption rates, and the degree of toxicity of the particular nanomaterials used. Risk assessments must be extremely accurate and trustworthy to detect and quantify the risks that the application of innovative materials could have in the food industry. Before these technologies are extensively employed as products in the agriculture and food sectors, thorough life cycle analyses, plant uptake of nanoparticles, bio-distribution, entry into the food chain, and other issues need to be investigated. Active nanomaterial biocidal substances must be evaluated and approved according to the rule. The public must be made aware of the advantages of adopting nanotechnology for other objectives than present technologies if the food and agriculture industries are to fully use its potential (Hossain et al., 2021).

1.4.1 EFFECT AND EXPOSURE OF NANOPARTICLES ON HUMAN HEALTH

Inhalation, ingestion, and dermal penetration are three ways that nanoparticles from artificial or other nanomaterials might enter the body. Nanoparticles may enter the body through medical implants and nanotechnology-based medical equipment and medications. Although the main exposure that should worry consumers is through ingesting, skin penetration, and inhalation are nearly entirely related to workers in the factories that produce nanomaterials. The major causes of nanoparticle presence in food are direct contact between food and nanopackaging and migration of nanoparticles from nanopackaging materials. The two main organs for the distribution of nanoparticles after ingestion and after they have passed from the intestines to the bloodstream are the liver and the spleen. The properties of nanoparticles have an impact on how they affect the human body. Particles that reach the bloodstream may have an impact on blood vessel function or lining, encourage the formation of blood clots, or have cardiovascular consequences related to breathing ambient ultrafine particles. The toxicity of nanoparticles relies on their characteristics, route of entry into the body, concentration, and exposure time, as well as on individual sensitivity and organismal state. Surface chemistry, agglomeration state, material source, production method, and storage should all be taken into consideration in this complex process in order to establish the physicochemical characteristics of nanomaterials and assess the toxicity of nanomaterials. Inflammation results from the activation of transcription factors that are responsive to oxidative stress, and this is significant because oxidative stress and chronic inflammation can result in a variety of particle-specific effects, including fibrosis, cancer, and genotoxicity brought on by fibers or secondary mutation. Among the potential negative impacts of nanoparticles, genotoxicity, and carcinogenicity are those that receive the most attention, aside from toxicity. Even though bulk ZnO is not harmful, ZnO nanoparticles have the potential to be genotoxic in human epidermal cells, which emphasizes the importance of particle diameter. According to some theories, the local production of granulomas and fibrosis in the lungs is what causes the carcinogenic consequences of persistent particles like asbestos. Additionally, prolonged inhalation of extremely high concentrations (10 mg/mL) of nano-TiO_2 has been linked to lung cancers (Sajid et al., 2015).

1.4.2 RISKS AND BENEFITS

According to individual data analysis, the significance of naturalness in food products and trust were determined to be significant factors influencing the perceived advantages and hazards of nanotechnology-based foods and food packaging. The general population is unfamiliar with the word "nanotechnology," according to previous surveys. The public perceives nanotechnology packaging as less troublesome. Consumers might be more open to packaging improvements than they are to food-related ones. The two factors nano-outside (such as packaging) and nano-inside were identified as the primary determinants of how each person perceives the use of nanotechnology in food (e.g., foods). According to the findings of several studies, nanotechnology packaging is viewed as being more advantageous and posing a lower risk to health than nanotechnology meals.

1.4.3 RISK AND SAFETY ISSUES

The fast advancement of nanotechnology has resulted in the formation of several scientific and industrial disciplines, with the market for nanomaterial-based goods constantly expanding. While nanotechnology has many benefits, its incorporation into the food business has prompted questions about public safety, environmental effects, ethical considerations, and regulatory frameworks. Nanomaterials can have different physicochemical and biological properties than their standard equivalents, and these undiscovered qualities may bring unexpected risks.

Even though the food sector has investigated the nanoencapsulation technology of bioactive chemicals in great detail, the potential danger of oral exposure to nanomaterials remains a concern. Depending on how susceptible they are to hydrolysis by digestive enzymes and the circumstances in the GIT, nanocarriers' destiny in the GIT varies dramatically (McClements and Xiao, 2017). Even so, unbound nanocarriers frequently pass through intestinal and cellular barriers, which may enhance the bioaccumulation of foreign substances in blood, tissues and cells of humans. The main cause of nanotoxicity is excessive exposure to reactive oxidant species, which causes *in vivo* cells to malfunction and lose the ability to perform their regular biological and physiological tasks. As a result, severe nanotoxicity can result in DNA mutation, unchecked cell signalling, changed cell motility, allergic lung inflammation, cytotoxicity, apoptosis, and fatal diseases like cancer.

Because of their toxicity, organic solvents and emulsifiers are frequently used in the manufacturing of nanocarriers, which can cause concerns. An evaporation procedure is required to remove the organic solvents, but it may result in unanticipated residual solvents remaining in the finished product, which could have safety problems if their concentration is unknown. Organizations like the Food and Drug Administration

(FDA), the World Health Organization (WHO), and the European Food Safety Authority (EFSA) have all identified safe usage levels for solvents and emulsifiers, despite the fact that they have classed these substances as harmful (McClements and Rao, 2011).

When exposed to human lung cells, TSilica nanoparticles employed as anti-caking agents can be harmful (Athinarayanan et al., 2014). Numerous elements, such as the particle's surface shape, concentration, surface energy, aggregation, and adsorption, have an impact on dissolution. Cushen et al. (2014) have created a model to investigate the migration of particles from food packaging. They investigated the migration of silver and copper from nanocomposites and found that, in comparison to contact duration, temperature, or particle size, the nanofiller percentage in the nanocomposites was one of the most important characteristics driving migration. As each nanomaterial is unique, toxicity will most likely be evaluated on a case-by-case basis. To assure product quality, health and safety requirements, and environmental laws, regulatory agencies must also create some criteria for commercial goods.

Food packaging must be constructed of inert materials for hygienic reasons, but increasingly, active and intelligent materials have also been sold. In addition to extending shelf life and improving storage conditions, enzymes, antibacterials, and absorbent substances can significantly simplify the distribution of food. These active particles with anti-microbial capabilities for packaging include manganese oxide, zinc oxide, and silver nanoparticles, for instance (Rhim et al., 2006). QD, carbon nanotubcs, gold nanostructures, and other active nanostructures have been or may be utilized as microbe sensors or in other food safety studies.

The digestive system has damaged some nourishment. The term "nutrition delivery system" refers to a device or nanocarrier that distributes food in precise locations. One of these systems is the chitosan nanocarrier. A significant component of the food industry is the extraction of nutrients from basic materials. New approaches are replacing traditional methods of food preparation due to the technology's quick proliferation. Without a doubt, nanotechnology will be a key factor in this advancement. Such methods in food processing may increase food processing yields and reduce waste or nutritional deterioration. Nutrition delivery systems must be composed of biodegradable material to avoid negative health effects on consumers.

1.5 FUTURE PERSPECTIVES OF NANOSCIENCE IN FOOD INDUSTRIES

While scientific research into the use of nanotechnology in the food sector has advanced quickly, research into nanotechnology related to nanostructures has advanced far more slowly. As the field of food nanotechnology research develops concerns concerning the safety of nanotechnology products for human consumption and use are growing.

Nanotechnology is gradually advancing in modern research because of its effectiveness, affordability, reproducibility, and of course, its small size. The food and agriculture industries are particularly delicate since they directly affect people's health. Nanosensors and nanobiosensors, which are extremely effective at detecting harmful levels, can play crucial roles in maintaining health safety in the modern world, which is highly concerned with production levels, health, quality, and environmental safety. We have looked at the various applications of nanotechnology in these significant fields, ranging from seed preservation to packaging.

The idea of "smart" packaging will probably spread in the future when antigen-specific markers will be employed to create thin films of polymer nanocomposites by adding nanoparticles to food packaging. The designated antigen markers may aid in the identification and isolation of bacteria that cause deterioration. Applications for new nanosystems should concentrate on becoming environmentally benign and minimizing their negative impacts on customers' food and the environment.

KEYWORDS

- **delivery system**
- **nanocapsule**
- **nanoemulsion**
- **nanotechnology**
- **sustainability**

REFERENCES

Athinarayanan, J., Periasamy, V. S., Alsaif, M. A., Al-Warthan, A. A., & Alshatwi, A. A. (2014). Presence of nanosilica (E551) in commercial food products: TNF-mediated oxidative stress and altered cell cycle progression in human lung fibroblast cells. *Cell Biology and Toxicology, 30*(2), 89–100. https://doi.org/10.1007/s10565-014-9271-8

Baldwin, E. A., Bai, J., Plotto, A., & Dea, S. (2011). Electronic noses and tongues: Applications for the food and pharmaceutical industries. *Sensors, 11*(5), 4744–4766. https://doi.org/10.3390/s110504744

Chaudhry, Q., Scotter, M., Blackburn, J., Ross, B., Boxall, A., Castle, L., Aitken, R., & Watkins, R. (2008). Applications and implications of nanotechnologies for the food sector. *Food Additives & Contaminants: Part A, 25*(3), 241–258. https://doi.org/10.1080/02652030701744538

Chen, C., Yuan, Z., Chang, H.-T., Lu, F., Li, Z., & Lu, C. (2016). Silver nanoclusters as fluorescent nanosensors for selective and sensitive nitrite detection. *Analytical Methods, 8*(12), 2628–2633. https://doi.org/10.1039/C6AY00214E

Couto, C., & Almeida, A. (2022). Metallic nanoparticles in the food sector: A mini-review. *Foods, 11*(3), 402. https://doi.org/10.3390/foods11030402

Cushen, M., Kerry, J., Morris, M., Cruz-Romero, M., & Cummins, E. (2014). Evaluation and simulation of silver and copper nanoparticle migration from polyethylene nanocomposites to food and an associated exposure assessment. *Journal of Agricultural and Food Chemistry, 62*(6), 1403–1411. https://doi.org/10.1021/jf404038y

Ding, W. K., & Shah, N. P. (2009). Effect of various encapsulating materials on the stability of probiotic bacteria. *Journal of Food Science, 74*(2), M100–M107. https://doi.org/10.1111/j.1750-3841.2009.01067.x

Gupta, R., & Tomer, A. K. (2017). Recent advances in the field of nanotechnology: A review. *IOSR Journal of Dental and Medical Sciences, 16*(01), 14–18. https://doi.org/10.9790/0853-1601111418

Honda, M., Asai, T., Oku, N., Araki, Y., Tanaka, M., & Ebihara, N. (2013). Liposomes and nanotechnology in drug development: Focus on ocular targets. *International Journal of Nanomedicine,* 495. https://doi.org/10.2147/IJN.S30725

Hossain, A., Skalicky, M., Brestic, M., Mahari, S., Kerry, R. G., Maitra, S., Sarkar, S., Saha, S., Bhadra, P., Popov, M., Islam, Mst. T., Hejnak, V., Vachova, P., Gaber, A., & Islam, T. (2021). Application of nanomaterials to ensure quality and nutritional safety of food. *Journal of Nanomaterials, 2021*, 1–19. https://doi.org/10.1155/2021/9336082

McClements, D. J., & Rao, J. (2011). Food-grade nanoemulsions: formulation, fabrication, properties, performance, biological fate, and potential toxicity. *Critical Reviews in Food Science and Nutrition, 51*(4), 285–330. https://doi.org/10.1080/10408398.2011.559558

McClements, D. J., & Xiao, H. (2017). Is nano safe in foods? Establishing the factors impacting the gastrointestinal fate and toxicity of organic and inorganic food-grade nanoparticles. *NPJ Science of Food, 1*(1), 6. https://doi.org/10.1038/s41538-017-0005-1

Men, H., Chen, D., Zhang, X., Liu, J., & Ning, K. (2014). Data fusion of electronic nose and electronic tongue for detection of mixed edible-oil. *Journal of Sensors, 2014*, 1–7. https://doi.org/10.1155/2014/840685

Mohareb, F., Papadopoulou, O., Panagou, E., Nychas, G.-J., & Bessant, C. (2016). Ensemble-based support vector machine classifiers as an efficient tool for quality assessment of beef fillets from electronic nose data. *Analytical Methods, 8*(18), 3711–3721. https://doi.org/10.1039/C6AY00147E

Rhim, J.-W., Hong, S.-I., Park, H.-M., & Ng, P. K. W. (2006). Preparation and characterization of chitosan-based nanocomposite films with antimicrobial activity. *Journal of Agricultural and Food Chemistry, 54*(16), 5814–5822. https://doi.org/10.1021/jf060658h

Sajid, M., Ilyas, M., Basheer, C., Tariq, M., Daud, M., Baig, N., & Shehzad, F. (2015). Impact of nanoparticles on human and environment: Review of toxicity factors, exposures, control strategies, and future prospects. *Environmental Science and Pollution Research, 22*(6), 4122–4143. https://doi.org/10.1007/s11356-014-3994-1

Scrinis, G., & Lyons, K. (2007). The emerging nano-corporate paradigm: Nanotechnology and the transformation of nature, food and agri-food systems. *The International Journal of Sociology of Agriculture and Food,* 22–44. https://doi.org/10.48416/IJSAF.V15I2.293

Sekhon, B. S. (2010). Food nanotechnology—An overview. *Nanotechnology, Science and Applications, 3*, 1–15.

Sundramoorthy, A. K., Vignesh Kumar, T. H., & Gunasekaran, S. (2018). Graphene-based nanosensors and smart food packaging systems for food safety and quality monitoring. In *Graphene bioelectronics* (pp. 267–306). Elsevier. https://doi.org/10.1016/B978-0-12-813349-1.00012-3

Torney, F., Trewyn, B. G., Lin, V. S.-Y., & Wang, K. (2007). Mesoporous silica nanoparticles deliver DNA and chemicals into plants. *Nature Nanotechnology, 2*(5), 295–300. https://doi.org/10.1038/nnano.2007.108

Voke, E., Pinals, R. L., Goh, N. S., & Landry, M. P. (2021). *In Planta* nanosensors: Understanding biocorona formation for functional design. *ACS Sensors, 6*(8), 2802–2814. https://doi.org/10.1021/acssensors.1c01159

Yu, G., Wu, W., Zhao, Q., Wei, X., & Lu, Q. (2015). Efficient immobilization of acetylcholinesterase onto amino functionalized carbon nanotubes for the fabrication of high sensitive organophosphorus pesticides biosensors. *Biosensors and Bioelectronics, 68*, 288–294. https://doi.org/10.1016/j.bios.2015.01.005

Yu, J., Jeon, Y.-R., Kim, Y.-H., Jung, E.-B., & Choi, S.-J. (2021). Characterization and determination of nanoparticles in commercial processed foods. *Foods, 10*(9), 2020. https://doi.org/10.3390/foods10092020

CHAPTER 2

Advancement of Nanotechnology in the Food Industry

ABHINAY THAKUR[1], HARPREET KAUR[1*], RAMESH CHAND THAKUR[2], and ASHISH KUMAR[3*]

[1]Department of Chemistry, School of Chemical Engineering and Physical Sciences, Lovely Professional University, Phagwara, Punjab, India

[2]Department of Chemistry, Himachal Pradesh University, Shimla, Himachal Pradesh, India

[3]NCE, Bihar Engineering University, Department of Science and Technology, Government of Bihar, India

*Corresponding author

ABSTRACT

The field of nanotechnology has witnessed significant advancements in past decades, and its applications in the food industry have garnered significant attention. This chapter focuses on the impact of nanotechnology on food production, packaging, preservation, and nutrition while examining the various ways it has transformed the food enterprise. The application of nanoparticles to food packaging has considerable potential. Nanoclays and nanocomposites have improved the barrier properties of packaging materials, protecting food from moisture, gases, and contaminants. Additionally, intelligent packaging equipped with nano-enabled sensors and indicators can monitor and communicate vital information about food freshness, temperature, and microbial contamination, ensuring consumer safety. Active packaging, incorporating nanomaterials with antimicrobial properties, has also developed as a remarkable tool for

Nanotechnological Advancements in the Food Industry.
Ishrat Majid, Bababode Adesegun Kehinde, and Sheikh Shafat Hussain (Eds.)

DOI: 10.1201/9781774919750-2

extending the shelf life of perishable foods. The chapter then explores the usage of nanosensors for quality control in the food industry. Nanosensors offer high sensitivity and selectivity in detecting contaminants, including pesticides, heavy metals, and pathogens, ensuring rapid and accurate quality control. Furthermore, the chapter highlights the application of nanoencapsulation for improved nutrition in food products. This technology holds great potential for fortifying food products with essential nutrients and promoting health and wellness.

2.1 INTRODUCTION

The food industry exhibits a vital path in meeting the global demand for safe, nutritious, and sustainable food. The demand for creative solutions to problems with food production, processing, and distribution is increasing along with the global population. Nanotechnology has gained momentum recently and has the ability to completely transform the food industry (Berekaa, 2015; Nurfatihah and Siddiquee, 2019; Wesley et al., 2014). By manipulating matter at the nanoscale, scientists and researchers have developed novel applications that enhance food quality, safety, and sustainability. Nanotechnology is the study and practise of manipulating and controlling objects and systems at the nanoscale level, which is typically 1–100 nm (Sahoo et al., 2021; Hannon et al., 2015). At this scale, materials exhibit unique properties and behaviors, which can be harnessed to overcome limitations in various industries, including food. Figure 2.1 displays the morphologies of 0–3D nanostructured materials.

The exquisite management that nanotechnology provides over a material's structure and characteristics allows for the creation of useful nanomaterials with enhanced functionality. Since nanotechnology has the ability to solve major problems and offer a wide range of advantages, it has attracted a lot of interest in the food industry. The goal of this chapter is to examine the developments in nanotechnology as they relate to the food industry, with a particular emphasis on how they affect food preparation, preservation, and nutrition. Food packaging is one of the main sectors where nanotechnology has made substantial progress (Chung et al., 2017; Rashidi and Khosravi-Darani, 2011). Maintaining food quality, prolonging shelf life, and minimizing spoilage all depend heavily on packaging. To enhance their barrier qualities, packaging materials have been infused with

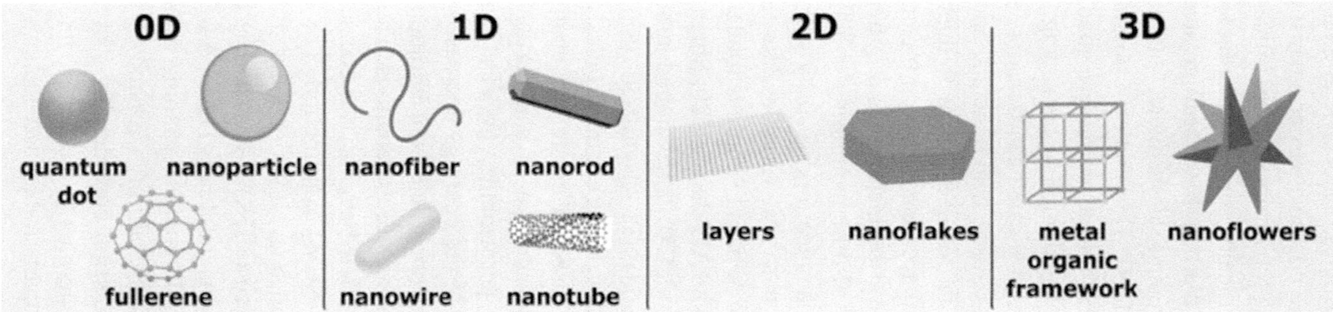

FIGURE 2.1 Illustration of some morphologies of 0–3D nanostructured materials.

Source: Reprinted from Bobrinetskiy et al. (2021). Copyright © 2021 by the authors. Licensee MDPI, Basel, Switzerland. https://creativecommons.org/licenses/by/4.0/

nanomaterials such nanoparticles, nanoclays, and nanocomposites. These nanomaterials create a more effective barrier against moisture, gases, and contaminants, protecting the food from external influences and ensuring its safety and freshness. Additionally, smart packaging systems incorporating nano-enabled sensors and markers have been created to track numerous factors in real-time, including temperature, humidity, and microbial contamination (Farhang, 2009). This makes it possible for prompt interventions and guarantees that consumers receive food that is both healthy and of the highest calibre. In addition to packaging, nanotechnology has made it possible to create sophisticated nanosensors for food sector quality control. Contaminants like viruses, heavy metals, and pesticides can seriously endanger human health. Nanosensors, with their high sensitivity and selectivity, offer a rapid and accurate detection method for such contaminants. These nanosensors can be included into systems for growing and processing food, offering real-time monitoring and allowing preventative measures to be taken to guarantee food security and purity. Nanosensors could also keep track of other vital variables like temperature, pH, and moisture content, which can be used to improve food processing conditions and reduce waste (Ghidan and Al Antary, 2020; Singh et al., 2017).

In an experiment, Li et al. (2017) developed antimicrobial packaging films using poly(lactic acid) (PLA) infused using titanium dioxide (TiO_2) nanoparticles and silver (Ag) nanoparticles. An approach called solvent volatilization was used to create the coatings. In comparison to pristine PLA sheets, the addition of nanoparticles resulted in lower opacity and water vapour permeability (WVP). Elastic modulus (EM) and Tensile strength (TS) of the coatings decreased whereas elongation at break rose with rising nanoparticle concentration in PLA, according to the mechanical characteristics of the coatings. The cross-section of the coatings made with nanoparticle blending was coarser, as shown by scanning electron microscopy (SEM) examination. The coatings made from nanoparticles did not establish any additional chemical bonds, according to Fourier-transform infrared (FTIR) research. The melting temperature (*T*m) and glass transition temperature (*T*g) of PLA were unaffected by the nanoparticles, according to differential scanning calorimetry (DSC) data, but they enhanced the cold crystallinity (*X*c) and crystallization temperature (Tc). Nanoparticle inclusion considerably increased the thermostability of the PLA sheets, according to thermogravimetric analysis (TGA). Additionally, the PLA nanoparticle-blended coatings demonstrated strong antibacterial action toward E. coli and Listeria monocytogenes. Migration studies, which are significant, showed that the discharge of nanoparticles from the coatings was within the acceptable range.

The employment of nanotechnology in the food industry has yielded significant global examples of advancements. Researchers at the Toronto university manufactured a nanotechnology-based food packaging material with embedded nanosensors that change color in the presence of harmful bacteria, providing a visual indication of food safety. This innovation has the potential to reduce foodborne illnesses and wastage. Additionally, a team at the University of Illinois created a portable nanosensor device capable of detecting allergens in food, enabling rapid and on-site testing to aid individuals with food allergies. Moreover, nanostructured lipid carriers developed at the University of Massachusetts protect fat-soluble vitamins during processing and enhance their bioavailability, potentially addressing nutrient deficiencies globally. Nanotechnology has also been applied to water purification, as seen in the nanoparticle-based system developed at Rice University, which effectively removes contaminants from water sources, crucial for food safety and production (Panpatte and Jhala, 2019). Researchers at the National University of Singapore have also extended the life span of perishable items through the invention of antimicrobial nanocoatings, lowering waste and improving food security. These instances highlight the various ways that nanotechnology is being used to advance the global food sector's sustainability, safety, and quality. Table 2.1 lists the benefits of using several types of nanoparticles in the food business (Thiruvengadam et al., 2018; Yu et al., 2018):

Another exciting area where nanotechnology has made substantial progress is in the domain of nanoencapsulation. Nanoencapsulation involves the encapsulation of bioactive compounds, nutrients, and flavors at the nanoscale. By improving these chemicals' stability, solubility, and accessibility, this method enables controlled release and increased efficacy. Functional foods, dietary supplements, and nutraceuticals now contain critical minerals and bioactive chemicals more effectively owing to nanoencapsulation. It offers opportunities to enhance the nutritional value and health benefits of food products, addressing public health concerns related to nutrient deficiencies and chronic diseases. While the advancements in nanotechnology in the food industry offer significant potential, it is important to consider safety and regulatory aspects (Biswas et al., 2022;

TABLE 2.1 The Utilization of Several Nanoparticles in the Food Industry and Their Advantages.

Nanoparticle	Application	Advantages
Silver nanoparticles	Food safety, packaging	• Strong antimicrobial properties, efficacious toward a broad spectrum of spoilage and pathogen-causing microorganisms. • Extended shelf life of food items.
Titanium dioxide nanoparticles	Food additives, packaging	• Enhances the opacity and whiteness of food products, used as a food colorant. • Provides UV protection, preserving the color and flavor of light-sensitive food.
Silicon dioxide nanoparticles	Food additives, packaging	• Acts as an anti-caking agent, improving flowability and preventing clumping of powdered food products. • Reduces moisture absorption while increasing the mechanical strength and barrier qualities of packaging materials.
Zinc oxide nanoparticles	Food additives	• Possesses antimicrobial properties, inhibiting the proliferation of bacteria and fungi in food items.

Bajpai et al., 2018). The sustainability of nanomaterials and their possible effects on health and the ecosystem must be carefully assessed, as with any newly developed technology. To create acceptable safety evaluation frameworks and recommendations for the use of nanomaterials in food applications, researchers and regulatory organizations are actively collaborating. By ensuring that nanotechnology is used ethically, these initiatives want to preserve customer trust in the food business. Future applications of nanotechnology in the food sector are quite promising. Emerging patterns show that this industry is still growing and innovating, as evidenced by the creation of nanobiosensors for quick disease identification, nanocoatings for antimicrobial surfaces, and nanomaterials for environmentally friendly food packaging. Figure 2.2 depicts the utilization of various techniques (voltammetry, potentiometry, electrochemical impedance and amperometry) in electrochemical biosensors for the detection of analytes and determination of their concentrations (Samal, 2017; Fahmy et al., 2020; Garde-Cerdán et al., 2021).

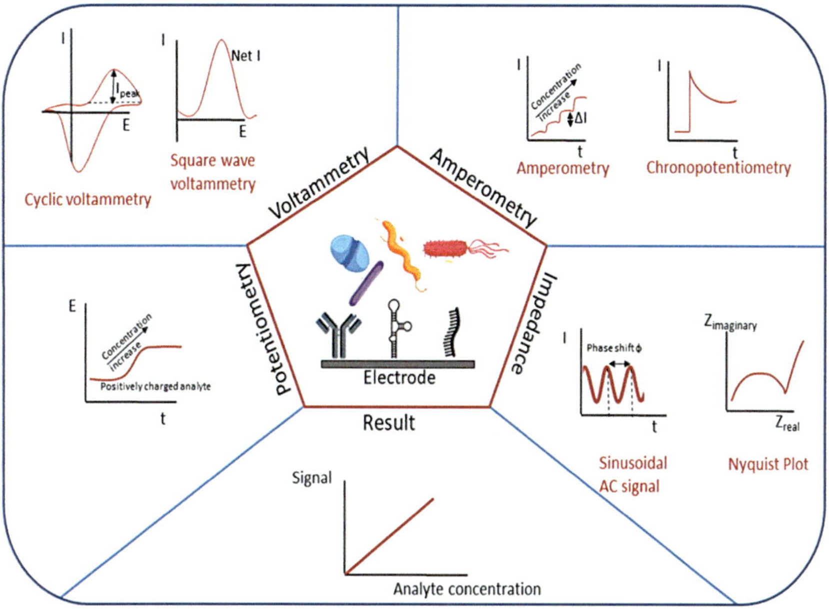

FIGURE 2.2 The utilization of various methods (voltammetry, potentiometry, electrochemical impedance, and amperometry) in electrochemical biosensors for the evaluation and detection of analyte concentration.

However, several challenges and ethical concerns need to be addressed. These include ensuring the scalability and cost-effectiveness of nanotechnology-based solutions, minimizing potential environmental impacts, and considering ethical implications related to consumer acceptance, labeling, and transparency.

2.2 IMPORTANCE OF ADVANCEMENTS IN NANOTECHNOLOGY

Nanotechnology, the manipulation and control of materials at the nanoscale, has brought about significant advancements in various industries, including the food industry. With its ability to alter the properties and behavior of materials, nanotechnology has opened up new possibilities for improving food production, packaging, preservation, and nutrition. In this section, we will provide insight into the advancements of nanotechnology in each of these sectors and explore the potential benefits and challenges associated with its implementation.

- **Food production:**

  By increasing crop yields, lowering the need for pesticides, and improving agricultural procedures, nanotechnology holds the ability to transform food production. For agricultural inputs like fertilizers and plant growth regulators, nanoparticles could be used as sophisticated delivery systems, ensuring their targeted and controlled release. This approach minimizes the environmental impact of agrochemicals and optimizes their utilization, leading to improved crop health and productivity (King et al., 2018). Nanosensors play a crucial role in monitoring and optimizing agricultural processes. These sensors can detect and measure various parameters, including soil moisture, nutrient levels, and pest infestation, providing real-time data for precision farming. By enabling farmers to make informed decisions based on accurate and timely information, nanosensors contribute to resource efficiency and sustainable agricultural practices.

- **Food packaging:**

  By strengthening the barrier qualities of packaging substances, strengthening food security, and prolonging shelf life, the usage of nanoparticles in food packaging has transformed the business (Siddiqui and Alrumman, 2021; Shah and Mraz, 2020; He and Hwang, 2016). Nanoclays, nanoparticles, and nanocomposites are incorporated into packaging materials to create a stronger and more effective barrier against moisture, gases, and contaminants. This ensures the integrity and freshness of the food throughout its shelf life and reduces the risk of spoilage and microbial growth. Intelligent packaging systems, equipped with nanosensors and indicators, enable real-time monitoring of various parameters, including temperature, humidity, and microbial contamination (Bashir et al., 2019; Dhonchak and Agnihotri, 2023; Kaya et al., 2023; Kumar and Thakur, 2020; Parveen et al., 2020; Sharma et al., 2014; Thakur and Kumar, 2021, 2022, 2023; Thakur et al., 2021, 2022a, 2022b, 2022c, 2022d, 2022e, 2023a, 2023b; Verma et al., 2023). These systems provide valuable details about the condition of the packaged food, allowing for timely interventions and ensuring consumer safety. Active packaging, utilizing nanomaterials with antimicrobial properties, further enhances food preservation by inhibiting the growth of spoilage-causing microorganisms (Nidhi, 2022; Kaur et al., 2020, 2022; Thakur et al., 2020) (Figure 2.3).

- **Food preservation:**

  Nanotechnology has introduced innovative solutions for food preservation, addressing the challenges associated with food spoilage, degradation, and waste. Nanostructured materials, such as nanoemulsions and nanocoatings, can be applied to food surfaces to create antimicrobial barriers, preventing the proliferation of pathogens and further spoilage inducing microorganisms. These nanomaterials have shown promise in enhancing the security and prolonging the shelf life of various food items, including fruits, vegetables, meats, and dairy (Kalpana Sastry et al., 2013; Gunasekaran, 2014; Mlalila et al., 2016). Furthermore, nanotechnology enables the development of controlled-release systems for antimicrobial agents and antioxidants. These active substances are enclosed using nanoencapsulation techniques, which prevents their deterioration and allows for their progressive release over time. This approach ensures the sustained efficacy of preservatives and antioxidants, contributing to the preservation of food quality and nutritional value.

- **Food nutrition:**

  Nanoencapsulation significantly enhances the distribution and bioavailability of nutrients and bioactive substances in food products. These substances can have their stability, solubility, and absorption improved by being nanoencapsulated. Bioactive substances can be released under control using nanoencapsulation procedures, ensuring that they are delivered to the body's desired locations. Nanotechnology also enables the fortification of food products with essential nutrients, addressing global nutrition challenges (Abbas et al., 2009; Subhan et al., 2021). Sensitive nutrients, including minerals and vitamins, could be encapsulated and protected using nanostructured delivery systems, improving

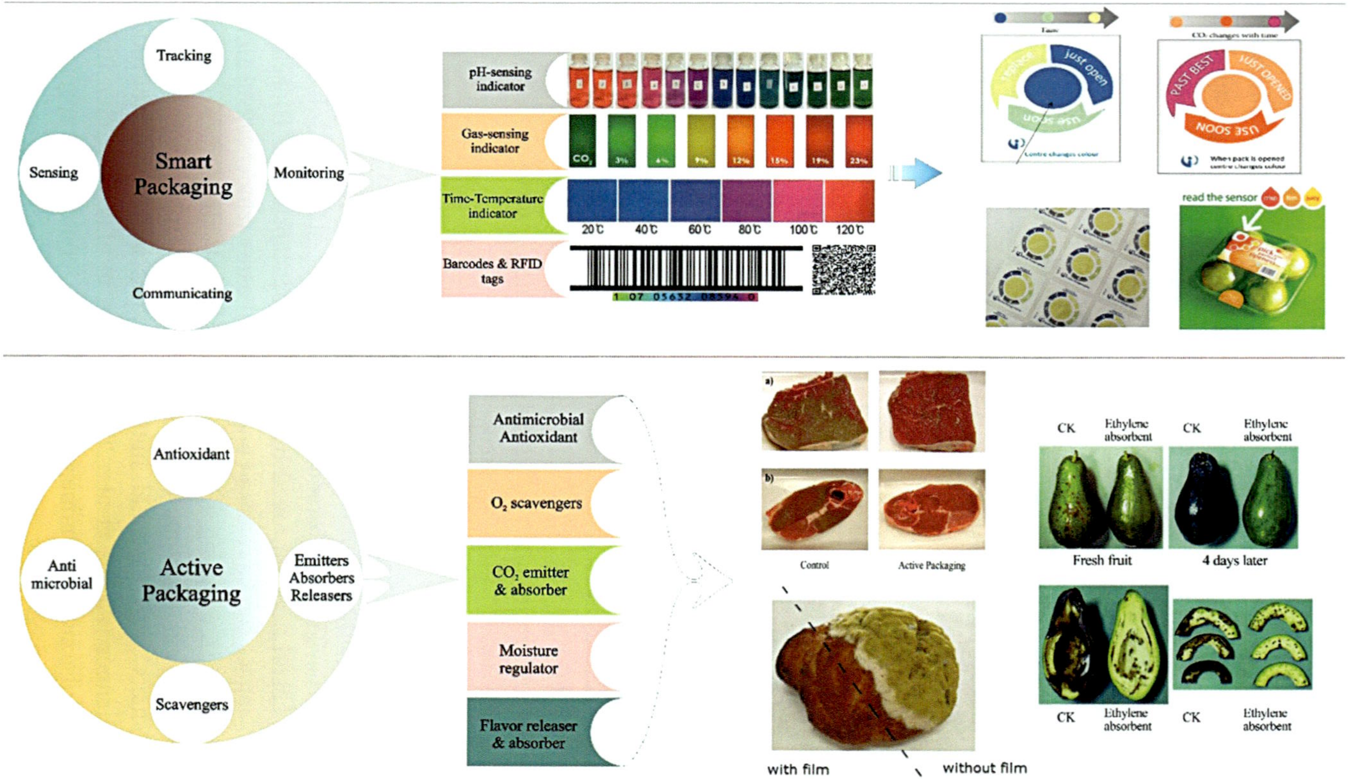

FIGURE 2.3 Characteristics, grouping, and use of active and intelligent packaging materials.

Source: Reprinted with permission from Sani et al. (2021). Copyright © 2021 by the authors. Licensee MDPI, Basel, Switzerland. https://creativecommons.org/licenses/by/4.0/

their durability and reducing deterioration throughout food preparation and preservation. This approach allows for the development of functional foods with improved nutritional profiles, targeting specific dietary needs and promoting better health outcomes.

2.3 NANOMATERIALS FOR FOOD PACKAGING

Nanotechnology has revolutionized the field of food packaging by introducing nanomaterials that offer enhanced barrier properties, intelligent packaging systems, and active packaging solutions for extended shelf life. These advancements have significantly improved food safety, preservation, and quality. In this section, we will explore each of these areas in detail.

2.3.1 ENHANCED BARRIER PROPERTIES

To preserve the integrity, freshness, and safety of food goods, packaging must have improved barrier qualities. In order to improve the barrier qualities of packaging substances, nanomaterials have emerged as a possible approach. They offer unique properties at the nanoscale, allowing for precise control over permeability to gases, liquids, and contaminants. The use of nanomaterials, such as nanoparticles, nanoclays, and nanocomposites, to enhance the barrier qualities of food packaging will be covered in more detail in this section. Layered nanomaterials called nanoclays, such as montmorillonite and halloysite, could be added to packaging substances to improve their barrier qualities. These nanoclays create tortuous pathways within the packaging matrix, reducing the permeability of gases and liquids. By effectively blocking the passage of molecules, nanoclays help sustain the quality and freshness of packaged food products. For example, nanoclays can be integrated into polymer matrices to form nanocomposite films. The shelf life of perishable goods is increased by these films' enhanced resistance to gases including oxygen, water vapor, and other flammable gases. Nanoparticles, such as silver nanoparticles, have also been extensively researched for their antibacterial characteristics in food packaging in addition to nanoclays. Due to their natural antibacterial properties, silver nanoparticles successfully prevent the development of germs and fungi that can cause food deterioration and foodborne illnesses. By incorporating silver nanoparticles into packaging films, a defensive barrier is created, mitigating the proliferation of microorganisms and maintaining the safety and freshness of the packaged food. The controlled release of silver ions from the nanoparticles further enhances the antimicrobial effect, ensuring long-term preservation (Bayda et al., 2020; Chellaram et al., 2014; Huang et al., 2018; Kus-liśkiewicz et al., 2021).

Furthermore, nanocomposites, which combine nanoparticles or nanoclays with polymers, offer tailored barrier properties for specific food products. For instance, in the case of fatty foods, nanoparticles can be dispersed within the packaging film to create a hydrophobic barrier. This barrier prevents the migration of moisture and oxygen, which can deteriorate the quality of the food product, such as causing rancidity in oils or affecting the texture of snacks. By controlling the nanoparticle dispersion and concentration, packaging materials can be customized to address the specific needs of different food products, providing optimal protection and preservation. Nanomaterials not only enhance the barrier properties of food packaging but also offer additional advantages (Gómez-Arribas et al., 2018; Puscaselu et al., 2020). Their nanoscale size provides a large substrate area-to-volume proportion, allowing for efficient interactions with gases, liquids, and contaminants. Moreover, their unique properties can be tailored to specific packaging requirements, ensuring optimal performance. The versatility of nanomaterials enables the development of packaging solutions that cater to the diverse needs of different food products, ranging from fresh produce to processed foods. The utilization of nanotechnology in food packaging offer gained global attention, with numerous examples showcasing its effectiveness in enhancing barrier properties (Lammari et al., 2020; Niculescu and Grumezescu, 2022). Beneath we have explored some data-based examples from around the world that highlight the impact of nanomaterials in food packaging:

- **United States:**

  In the United States, researchers at the University of California, Davis, developed a nanocomposite film using cellulose nanocrystals derived from renewable resources. This film exhibited remarkable barrier attributes against water and oxygen vapor, significantly extending the shelf life of packaged fruits and vegetables. The nanocomposite film reduced the spoilage rate of strawberries by up to 65% compared to conventional packaging materials (Bobrinetskiy et al., 2021).

- **Europe:**

  In Europe, the EU-funded NanoPack project aimed to develop and commercialize antimicrobial nanocomposite films for food packaging. The project collaborated with partners from various countries, including Germany, the Netherlands, and Spain. NanoPack successfully developed a film containing natural clay nanotubes and essential oils that exhibited strong antimicrobial activity against foodborne pathogens (Sani et al., 2021; Pandey et al., 2022; Figueroa-Lopez et al., 2020). The nanocomposite film was used for packaging meat and cheese products, resulting in a significant reduction in bacterial growth and improved food safety.

- **Australia:**

  Researchers at the University of New South Wales in Australia developed a nanocomposite film with enhanced barrier properties against moisture and oxygen (Abdullahi and Nura, 2018; Cicek and Nadaroglu, 2015; Phata, 2010; Ummi and Siddiquee, 2019). The film incorporated nanoclay particles, which created a tortuous pathway that hindered the permeation of gases and liquids. The nanocomposite film was tested with meat products, demonstrating a substantial reduction in oxygen transmission rate and extended shelf life.

- **China:**

  In China, scientists at Zhejiang University developed a nano-composite film using graphene oxide and cellulose nanofibrils. The film exhibited excellent barrier properties against water and oxygen vapor, preserving the integrity of packaged fruits. In a study, strawberries packaged with the nanocomposite film showed a significantly reduced weight loss and retained their visual and sensory qualities for a longer duration compared to conventional packaging materials (Dasgupta et al., 2015; Roopa et al., 2022; Yadav et al., 2020).

- **South America:**

  In South America, researchers at the Federal University of Rio Grande do Sul in Brazil developed a nanocomposite film using nanocellulose fibers. The film showed improved barrier qualities against oxygen and water vapour, extending the shelf life of packaged baked goods. For improved product quality for consumers, the nanocomposite layer efficiently reduced moisture absorption and prevented bread from staling. It is important to note that the use of nanomaterials in food packaging raises questions about how they can end up in the food and how that might affect people's health (Dhewa, 2015; Mura et al., 2013; Sozer and Kokini, 2009). Nevertheless, a lot of study is being done to make sure that nanoparticles used in food packaging are safe. The safety of nanomaterials and their potential for migration has been evaluated by regulatory agencies through the implementation of guidelines and regulations. Research is being done to determine safe consumption levels and to comprehend the potential dangers linked with nanomaterials.

2.3.2 INTELLIGENT PACKAGING SYSTEMS

Intelligent packaging systems, enabled by nanotechnology, have revolutionized the traditional concept of food packaging by incorporating advanced features that go beyond passive protection. These systems integrate nanosensors, indicators, and smart labels to offer real-time details about the condition and quality of packaged food, enhancing consumer trust and convenience. Nanosensors embedded within packaging materials act as intelligent detectors, capable of monitoring various parameters that affect food quality, such as temperature, humidity, and gas composition. Nanosensors offer high selectivity and sensitivity, enabling precise monitoring and detection of specific gases or compounds. For instance, carbon nanotubes can be employed to create nanosensors that detect ethylene gas, a ripening hormone emitted by fruits. By continuously monitoring ethylene levels, these nanosensors can provide valuable information about the ripening stage of fruits, enabling timely interventions such as adjusting storage conditions or removing ripe fruits to prevent over-ripening and spoilage. This real-time monitoring allows for better inventory management, reduces food waste, and ensures that consumers receive fresh and high-quality products (Dąbrowska et al., 2018). Time-temperature indicators (TTIs) are another component of intelligent packaging systems.

TTIs are designed to change color or display visual signals when the product has been subjected to unfavorable conditions, including high temperatures during transportation or storage. These indicators offer a precise indication of potential quality deterioration, allowing consumers and retailers to make informed decisions regarding the safety and freshness of the product. By providing visual cues that are easy to interpret, TTIs empower consumers to assess product quality without relying solely on expiration dates or subjective judgments, promoting the safety of food and reducing the risk of consuming spoiled or unsafe food. Smart labels embedded with nanotechnology provide consumers with detailed information about the product, its origin, nutritional content, allergens, and storage instructions (Akgönüllü et al., 2021; Lemes et al., 2016; Ravichandran, 2010). These labels can be scanned using smartphones or other devices, granting consumers easy access to relevant information. By simply scanning the smart label, consumers can obtain information about the product's sourcing, processing methods, and any potential allergens, allowing them to make informed choices based on their dietary preferences, restrictions, or ethical considerations. Smart labels also enable consumers to verify the authenticity and traceability of the product, enhancing transparency and promoting consumer confidence.

Systems for intelligent packaging help to make the food supply chain more effective and sustainable overall. These systems enable preemptive actions to be taken to prevent quality deterioration, reduce food waste, and enhance inventory management by delivering real-time information about the status of the product. Moreover, the enhanced transparency and accessibility of information through smart labels promote a more informed and responsible consumer base, fostering a culture of conscious consumption and sustainable choices (Kulkarni et al., 2022).

Intelligent packaging systems incorporating nanotechnology have gained global recognition for their ability to provide real-time information and enhance consumer confidence in food products. Here are some data-based examples from around the world that showcase the implementation and impact of intelligent packaging systems:

- **United States:**

  Using nanotechnology, the American company FreshTag has created intelligent time-temperature indicators (TTIs). These indicators are applied to food packaging and change color based on the time and temperature conditions experienced by the product. Consumers can easily interpret the color change to assess the freshness and safety of the food. A study conducted by the University of Florida demonstrated that the use of intelligent TTIs significantly reduced food waste by improving consumer awareness and reducing the discard of still safe-to-eat food.

- **Europe:**

The European Union-funded NanoPack project in Europe sought to create intelligent packaging solutions to increase the lifespan of perishable goods. The project successfully developed active nanocomposite films embedded with antimicrobial agents and TTIs. These films were tested on various food products, including bread and cheese, and demonstrated improved preservation and freshness. The project estimated that the implementation of intelligent packaging could potentially reduce food waste by up to 20% (Singh et al., 2021).

- **Australia:**

Researchers at the University of Melbourne in Australia developed an intelligent packaging system using nanosensors to monitor the freshness of meat products. The nanosensors, embedded in the packaging material, detected volatile organic compounds (VOCs) produced during meat spoilage (Maluin et al., 2021; Sridhar et al., 2021). The data collected by the sensors was transmitted wirelessly to a smartphone application, offering real-time details to consumers about the freshness and safety of the meat. This innovative packaging system has the potential to minimize foodborne illnesses and reduce food waste by enabling consumers to make informed decisions.

- **South Korea:**

In South Korea, a company called C-fresh developed intelligent packaging for fruits and vegetables. The packaging integrated nanosensors that monitored the levels of ethylene gas, which affects the ripening process. By continuously monitoring ethylene levels, the packaging system provided real-time information to retailers, enabling them to optimize storage conditions and reduce spoilage.

As a result, the company reported a significant reduction in food waste and improved product quality.

- **India:**

  In India, researchers at the Indian Institute of Technology (IIT) in Mumbai developed an intelligent packaging system using nanosensors to monitor the purity of milk. The nanosensors detected alterations in pH and temperature, which are indicators of milk spoilage. The data collected by the sensors was wirelessly transmitted to a smartphone application, allowing consumers to assess the quality of the milk before consumption. The implementation of this intelligent packaging system has the potential to reduce the consumption of spoiled milk and improve consumer health.

However, the implementation of intelligent packaging systems does come with challenges. The integration of nanosensors, indicators, and smart labels into packaging materials requires careful consideration of their compatibility, stability, and safety. Regulations and guidelines are being developed to ensure the safe and responsible usage of nanotechnology in food packaging. Additionally, there is a need for standardized communication platforms and technologies to ensure seamless compatibility between smart labels and consumer devices.

2.3.3 ACTIVE PACKAGING

Active packaging made possible through the integration of nanotechnology, has revolutionized the concept of extending the shelf life of food items. Nanotechnology allows for the regulated discharge of antimicrobials, oxygen scavengers, and other active compounds. This improves food preservation and lowers waste by inserting active agents into packaging materials. Antimicrobial nanoparticles or nanocomposites are key components of active packaging. These nanomaterials release antimicrobial agents, such as silver ions, which have potent inhibitory effects on spoilage-causing microorganisms. Perishable food items, like fruits, vegetables, and meats, have a longer shelf life because of the regulated release of these chemicals, which also helps prevent microbial growth. Numerous investigations have shown how silver nanoparticles can improve food safety by minimizing microbial contamination (Fotie et al., 2020; Maluin et al., 2020).

Oxygen scavenging is another crucial aspect of active packaging. Oxygen plays a vitality in food spoilage, as it promotes oxidative reactions that lead to the degradation of color, flavor, and texture. Nanomaterials, such as iron nanoparticles, can be incorporated into packaging films to scavenge oxygen molecules, preventing their contact with the food and delaying the onset of oxidation. This oxygen scavenging capability helps maintain the freshness and quality of food products, especially those prone to oxidation, such as snacks, oils, and bakery goods. Nanotechnology also enables the integration of other active agents into packaging materials to address specific food preservation challenges. For instance, antioxidants can be incorporated to inhibit the oxidation of fats and oils, extending their shelf life. Moisture regulators, such as zeolites, can help control humidity levels within the packaging, preventing moisture-related deterioration. Flavor enhancers can be encapsulated in nanocarriers and integrated into packaging materials to release pleasant aromas and enhance sensory attributes. The benefits of active packaging solutions based on nanotechnology are numerous. First and foremost, they greatly increase the shelf life of food items, cutting down on food waste and promoting sustainability. Active packaging lessens the need to throw away food too soon because it keeps it fresh and high-quality for longer periods of time. This, in turn, has positive environmental and economic implications. Furthermore, active packaging enhances food safety by inhibiting the proliferation of spoilage-causing microorganisms and reducing the risk of foodborne illnesses. The controlled release of antimicrobial agents provides an additional layer of protection against microbial contamination, ensuring that food products remain safe and suitable for consumption. Implementations of active packaging solutions incorporating nanotechnology have been witnessed worldwide, demonstrating their efficacy in extending the shelf life of food items. Here are some global examples showcasing the utilization and impact of active packaging on a global scale:

- **United States:**

  In order to increase the shelf life of fresh meats and seafood, a company called BluWrap in the US has created an active packaging method that uses a naturally occurring gas. The packaging technology regulates the carbon dioxide and oxygen levels around the food, which significantly reduces microbial development and food deterioration. This innovative active packaging system has enabled the transportation of fresh seafood from Alaska to various destinations, significantly reducing spoilage and extending product availability.

- **Europe:**

  In Europe, the NanoBarrier project, funded by the European Union, aimed to develop active packaging solutions using nanomaterials to extend the shelf life of food. The project focused on integrating antimicrobial agents and oxygen scavengers into packaging materials. The developed active packaging films demonstrated remarkable efficacy in inhibiting microbial growth and preserving food quality. The project's findings showcased the efficacy of nanotechnology-based active packaging in reducing food waste and enhancing sustainability efforts.

- **Australia:**

  Active packaging for prolonging the shelf life of fruits has advanced in Australia. Researchers at the University of New South Wales developed an active packaging film incorporating silver nanoparticles to combat fungal decay in strawberries. The silver nanoparticles exhibited strong antifungal properties, effectively inhibiting the proliferation of fungal pathogens and extending the shelf life of strawberries by several days. This technology offers promising opportunities to reduce post-harvest losses and improve fruit quality.

- **China:**

  In China, researchers have explored the viability of active packaging solutions for meat preservation. A study conducted by scientists from Jiangnan University developed a nanocomposite film incorporating antimicrobial nanoparticles for meat packaging. The active packaging film displayed effective microbiological growth prevention, enhanced mechanical strength, and improved water vapor barrier qualities, prolonging the shelf life of meat products. This innovation has the potential to significantly reduce meat spoilage and enhance food safety.

- **South Africa:**

  In South Africa, researchers at Stellenbosch University have developed an active packaging system to combat the ripening and deterioration of avocados. The system incorporates an ethylene scavenger into the packaging material, reducing the exposure of avocados to the ripening hormone ethylene. The shelf life of avocados has been extended and post-harvest losses have decreased owing to this active packaging method, which has demonstrated encouraging results (Mohammad et al., 2022; Musial et al., 2020; Primožiˇc et al., 2021; Wróblewska-Krepsztul et al., 2019). Additionally, active packaging can improve consumer satisfaction by preserving the nutritional value, texture, and sensory characteristics of the packaged food. The use of antioxidants, moisture regulators, and flavor enhancers helps maintain the desired characteristics of the food, enhancing its appeal and overall consumer experience. It is important to note that the development and implementation of active packaging solutions require thorough evaluation of their safety, efficacy, and regulatory compliance. Researchers and industry professionals are actively working toward ensuring that these technologies meet stringent standards and guidelines to ensure consumer health and confidence (Elemike et al., 2019; Kraśniewska et al., 2020; Li et al., 2017).

2.4 NANOSENSORS FOR QUALITY CONTROL

2.4.1 DETECTION OF CONTAMINANTS

Ensuring the safety and quality of food products is a critical concern in the food industry. Contamination by various substances, including pathogens, pesticides, heavy metals, and toxins, can pose significant risks to consumers' health. Detecting and monitoring these contaminants is essential to prevent foodborne illnesses, protect public health, and maintain consumer confidence. Nanosensors have emerged as powerful tools in the detection and monitoring of contaminants in the food supply chain due to their high sensitivity and selectivity. Nanosensors offer several advantages over traditional detection methods. They provide faster and more accurate results, enabling early identification and mitigation of potential risks. These sensors can be designed to specifically target and detect contaminants at very low concentrations, enhancing the sensitivity of detection and improving the overall effectiveness of quality control processes.

By incorporating nanosensors into the food industry's quality control procedures, food safety can be enhanced, and potential health risks can be minimized (Barhoum et al., 2022; Hano and Abbasi, 2022; Martirosyan and Schneider, 2014). One area where nanosensors have made significant contributions is in the detection of microbial contaminants. Pathogenic microorganisms, such as Salmonella, Escherichia coli, and Listeria, are major concerns as they can cause foodborne illnesses and outbreaks. Nanosensors can utilize specific recognition elements, such as antibodies or aptamers, to detect and quantify these pathogens with high sensitivity. The use of nanosensors, such as graphene-based nanosensors, has shown remarkable performance in rapidly and reliably detecting bacterial contaminants in food samples. These nanosensors can provide real-time results, allowing for prompt actions to be taken to prevent the spread of pathogens and ensure the safety of the food supply (Harish et al., 2022; Raja et al., 2021; Shafiq et al., 2020). Nanosensors are also valuable in detecting chemical contaminants, including pesticides and mycotoxins. Pesticide residues in fruits, vegetables, and grains can have adverse effects on human health. Nanosensors can detect and quantify these residues with high precision, providing accurate information for determining compliance with safety standards. Similarly, mycotoxins produced by fungi in food products, such as aflatoxins, can be detected using nanosensors. Early identification and prevention of the presence of mycotoxins are crucial to minimize health risks associated with contaminated food. Nanosensors offer the sensitivity and selectivity required for precise detection, enabling effective control measures to be implemented.

Heavy metal contamination is another significant concern in the food industry. Nanosensors incorporating nanostructured materials, such as nanowires and nanotubes, can selectively detect and quantify heavy metals, including lead, mercury, and cadmium. These sensors possess high sensitivity and selectivity, allowing for accurate monitoring of heavy metal levels in food samples. By utilizing nanosensors, food manufacturers and regulatory bodies can ensure compliance with safety regulations and minimize the risks associated with heavy metal contamination. In summary, nanosensors have proven to be powerful tools for the detection of contaminants in the food industry. Their high sensitivity, selectivity, and ability to detect contaminants at low concentrations make them invaluable assets in ensuring food safety and quality. Ensuring the safety and quality of food products is a global concern, and the use of nanosensors has gained significant attention in addressing this issue. Let's explore some global data examples that highlight the importance and impact of nanosensors in food safety and quality control:

- **Pathogen detection:** According to the World Health Organization (WHO), foodborne illnesses affect millions of people worldwide each year. Nanosensors have demonstrated their effectiveness in detecting pathogenic microorganisms and preventing foodborne outbreaks. For instance, a study conducted in China utilized nanosensors to detect Salmonella in poultry products, leading to improved food safety practices and reduced health risks for consumers (Naresh and Lee, 2021; Shang et al., 2019).
- **Pesticide residue monitoring:** The presence of pesticide residues in food can have detrimental effects on human health. The European Food Safety Authority (EFSA) reported that approximately 4% of tested food samples contained pesticide residues above the legal limits. Nanosensors offer precise and sensitive detection of pesticide residues, allowing for more accurate monitoring and compliance with safety regulations.
- **Mycotoxin control:** Mycotoxins, produced by fungi in food products, pose a significant risk to human health. The Food and Agriculture Organization (FAO) estimates that mycotoxins contaminate around 25% of global food crops (Neethirajan et al., 2018; Weiss et al., 2006). Nanosensors have been utilized to detect mycotoxins, such as aflatoxins, in various food commodities, enabling early identification and prevention of their presence.
- **Heavy metal monitoring:** Heavy metal contamination, including lead, mercury, and cadmium, is a global concern. The United Nations Environment Programme (UNEP) highlights the adverse health effects of heavy metal exposure, particularly in children. Nanosensors incorporating nanostructured materials have shown great potential in detecting and quantifying heavy metals in food samples, contributing to improved monitoring and mitigation strategies.
- **Global regulations and standards:** International organizations and regulatory bodies have recognized the significance of nanosensors in ensuring food safety and quality. For example, the U.S. Food and

Drug Administration (FDA) has established guidelines for the use of nanotechnology in food-related applications, emphasizing the importance of accurate detection methods. The European Union (EU) has also implemented regulations governing nanomaterials used in food contact materials, highlighting the need for robust quality control measures. These global data examples illustrate the widespread recognition of nanosensors as valuable tools for food safety and quality control. The adoption of nanosensor technology in various regions is driven by the aim to mitigate health risks, reduce foodborne illnesses, and enhance consumer trust in the food supply chain.

By integrating nanosensors into quality control processes, the food industry can enhance its ability to detect and monitor contaminants, mitigate risks, and protect consumer health. Continued research and development in nanosensor technology will further advance the field, enabling more effective and efficient food safety measures.

2.4.2 REAL-TIME MONITORING

Real-time monitoring of food parameters is a critical aspect of ensuring the quality, safety, and freshness of food products throughout the supply chain. Traditional methods of monitoring food parameters, such as temperature and humidity, often rely on manual measurements and periodic inspections. However, these methods are limited in their ability to provide timely and continuous data, leaving room for potential quality and safety issues to arise unnoticed. Nanosensors offer a transformative solution by providing continuous, real-time monitoring of food parameters, enabling proactive measures to be taken when necessary. Temperature is one of the most important factors affecting the quality and safety of food products. Proper temperature control is essential to prevent the growth of pathogenic bacteria, enzyme activity, and chemical reactions that can lead to spoilage and the degradation of nutritional value. Nanosensors can be integrated into packaging materials or directly into food products to monitor temperature changes during storage and transportation. These sensors can provide instant feedback on temperature variations, allowing for prompt action to be taken in case of temperature excursions that may compromise product quality or safety.

For example, nanosensors based on thermochromic materials can change color at specific temperature thresholds, providing visual indicators of temperature abuse. This real-time monitoring capability ensures that food products are maintained within safe temperature ranges, reducing the risk of spoilage and enhancing consumer safety.

Humidity is another crucial parameter that significantly impacts food quality and shelf life. Excessive moisture can lead to microbial growth and spoilage, while inadequate moisture can cause food products to dry out and lose their desired texture and flavor. Nanosensors offer the ability to accurately monitor humidity levels in packaging materials or storage environments, ensuring optimal conditions for food preservation. These sensors can detect even small changes in humidity and provide continuous data on moisture levels. By maintaining appropriate humidity levels, nanosensors help prevent quality deterioration and extend the shelf life of food products. This is particularly important for perishable items such as fruits, vegetables, and baked goods that are highly sensitive to moisture. In addition to temperature and humidity, nanosensors can also monitor other important parameters that affect food quality and safety. pH, for example, plays a vital role in determining the safety and quality of food products, as it affects microbial growth, enzyme activity, and overall product stability. Nanosensors integrated into packaging materials can monitor pH changes, providing insights into the deterioration of perishable products. These sensors can detect shifts in pH levels, indicating potential spoilage or changes in product quality. This real-time monitoring allows for timely intervention, such as removing or replacing compromised products, and helps minimize the risk of foodborne illnesses. Gas composition is another critical parameter that can be monitored in real-time using nanosensors. Certain gases, such as oxygen and carbon dioxide, are important indicators of food spoilage. Nanosensors based on nanomaterials can detect and quantify these gases, providing continuous data on gas composition within the packaging or storage environment. By monitoring gas levels, these sensors can provide early warnings of potential spoilage and allow for corrective actions to be taken, such as adjusting the packaging atmosphere or implementing preservative measures. This real-time gas monitoring capability helps extend the shelf life of food products and reduces the risk of consumers purchasing spoiled items. Furthermore, nanosensors can be used as freshness indicators by detecting specific volatile organic compounds (VOCs) emitted by food products as they deteriorate. These VOC sensors can be integrated into packaging materials or embedded in smart labels, allowing consumers and retailers to assess the freshness of a product by simply scanning it. This real-time freshness monitoring enables informed decision-making regarding product quality and helps reduce food waste by ensuring that products are consumed at their peak freshness.

Real-time monitoring of food parameters using nanosensors enhances the ability to maintain product quality and safety throughout the supply chain. By providing continuous feedback and early detection of deviations from desired conditions, nanosensors enable timely intervention, reducing the risk of food waste and ensuring consumer satisfaction. Moreover, the integration of nanosensors into smart packaging systems can enable real-time communication and data sharing, allowing stakeholders at various stages of the supply chain to access critical information about the status of food products. This information can be used to optimize storage conditions, prevent quality degradation, and make informed decisions regarding product distribution and shelf life. The use of nanosensors for real-time monitoring of food parameters is gaining traction in the food industry. However, several challenges still need to be addressed for their widespread implementation. These challenges include sensor calibration and validation, ensuring sensor stability and longevity, cost-effectiveness, and regulatory compliance. Ongoing research and development efforts are focused on addressing these challenges and further improving the performance and applicability of nanosensors in food monitoring applications (Sozer and Kokini, 2009).

2.5 NANOENCAPSULATION FOR IMPROVED NUTRITION

2.5.1 INTRODUCTION TO NANOENCAPSULATION

Nanoencapsulation is a technique that involves the encapsulation of bioactive compounds at the nanoscale to improve their stability, solubility, bioavailability, and controlled release. Bioactive compounds refer to substances that have a positive effect on health and are typically found in foods or natural sources. Examples of bioactive compounds include vitamins, minerals, antioxidants, polyphenols, essential fatty acids, and phytochemicals. The encapsulation process entails the formation of a protective shell or matrix around the bioactive compound, creating a nano-sized particle or carrier. The encapsulation materials can be classified into organic and inorganic materials, such as lipids, proteins, polysaccharides, and synthetic polymers (Ahmad, 2019a; Qu et al., 2021). These materials provide a physical barrier that shields the bioactive compound from external factors that may cause degradation, such as oxygen, light, heat, and moisture. The size range of the encapsulated particles in nanoencapsulation is typically between 1 and 100 nm. This size range offers several advantages, including a large surface area-to-volume ratio, increased stability, and improved solubility. The small particle size also allows for enhanced interactions with biological systems, such as improved absorption in the gastrointestinal tract and targeted delivery to specific sites in the body.

- **Food and beverages:**

  Nanoencapsulation has been extensively explored in the food and beverage industry to enhance the stability and bioavailability of bioactive compounds. For instance, in a study conducted in Australia, researchers encapsulated omega-3 fatty acids within nanoliposomes to improve their stability and incorporation into food products. The nanoencapsulated omega-3 fatty acids exhibited increased resistance to oxidation, allowing for the fortification of various food items, such as dairy products, bakery goods, and beverages, with these essential nutrients. In India, researchers used nanoencapsulation to improve the stability and bioavailability of curcumin, a bioactive compound found in turmeric. Encapsulating curcumin within nanocarriers enhanced its solubility and protected it from degradation. The resulting curcumin nanoformulations were incorporated into functional food products, such as beverages and dairy-based items, to enhance their health-promoting properties.

- **Pharmaceutical and nutraceutical industries:**

  Nanoencapsulation has revolutionized drug delivery systems by improving the bioavailability and targeted delivery of therapeutic compounds. In the United States, nanotechnology-based drug delivery platforms have been developed to encapsulate and deliver anticancer drugs. These nanoencapsulated drugs improve their efficacy, reduce side effects, and enable targeted delivery to tumor cells. In the nutraceutical industry, nanoencapsulation has been employed to enhance the bioavailability of dietary supplements and functional ingredients (Ahmad, 2020; Georgin et al., 2022; Liu et al., 2020). In Europe, researchers encapsulated vitamins, such as vitamin D and vitamin E, within nanocarriers to improve their stability and absorption. The nanoencapsulated vitamins were incorporated into nutraceutical formulations, including tablets, capsules, and powders, to enhance their bioavailability and health benefits.

- **Agricultural and plant sciences:**

  Nanoencapsulation has also found applications in agriculture and plant sciences to improve crop yield, nutrient absorption, and plant protection. In Brazil, researchers developed nanoencapsulated formulations containing micronutrients, such as iron and zinc, to overcome nutrient deficiencies in crops. The nanoencapsulated nutrients demonstrated improved stability and controlled release, leading to enhanced nutrient absorption by plants and increased crop productivity. Additionally, nanoencapsulation has been explored as a delivery system for agrochemicals, such as pesticides and herbicides. In China, scientists developed nanoencapsulated formulations of herbicides to reduce their environmental impact and improve their efficacy. The nanocarriers provided controlled release and targeted delivery of the herbicides, resulting in enhanced weed control and reduced chemical usage.

- **Cosmetics and personal care products:**

  Nanoencapsulation has gained popularity in the cosmetic industry due to its ability to improve the stability and efficacy of active ingredients. In South Korea, researchers encapsulated skin-brightening agents, such as kojic acid and arbutin, within nanocarriers to enhance their skin penetration and stability. The resulting nanoencapsulated formulations were incorporated into various cosmetic products, including creams, lotions, and serums, to improve their skin-lightening effects. Furthermore, nanoencapsulation has been utilized for the delivery of sunscreen agents. In Australia, scientists developed nanoencapsulated sunscreen formulations to enhance the photostability and efficacy of UV filters. The nanoencapsulated sunscreens offered improved protection against UV radiation, ensuring better sunscreen performance and prolonged sun protection.

2.5.2 ENHANCING STABILITY AND SOLUBILITY OF BIOACTIVE COMPOUNDS

Enhancing the stability of bioactive compounds through nanoencapsulation is crucial for maintaining their beneficial properties throughout food processing, storage, and consumption. The protective shell or matrix formed around the bioactive compound acts as a physical barrier, preventing direct contact with factors that can cause degradation, such as oxygen, light, heat, and moisture. For example, vitamins, such as vitamin C and vitamin E, are highly susceptible to degradation when exposed to oxygen and light. Nanoencapsulation can protect these vitamins by encapsulating them within a lipid-based matrix, shielding them from the detrimental effects of oxidation and photochemical reactions. This ensures that the vitamins remain stable and retain their potency, even in products with extended shelf lives.

Similarly, polyphenols, which are widely recognized for their antioxidant properties, are prone to degradation due to their chemical structure and sensitivity to environmental conditions. Nanoencapsulation can provide a protective environment for polyphenols, preventing their degradation caused by exposure to heat, light, and oxygen. This preservation of polyphenols' stability ensures that the bioactive compounds can exert their antioxidant effects and contribute to the overall health benefits of food products. In addition to stability enhancement, nanoencapsulation can significantly improve the solubility of bioactive compounds. Many bioactive compounds, such as curcumin and resveratrol, exhibit poor water solubility, limiting their incorporation into aqueous-based food systems. Nanoencapsulation addresses this challenge by reducing the particle size to the nanoscale, thereby increasing the surface area available for interaction with the solvent. The smaller particle size of the nanoencapsulated bioactive compounds enhances their dispersibility and solubility, enabling better incorporation into a wider range of food and beverage products. This improved solubility facilitates the uniform distribution of the bioactive compounds, ensuring their availability and bioavailability upon consumption. Moreover, the enhanced solubility allows for the formulation of clear, transparent products that are visually appealing and have improved sensory attributes. Nanoencapsulation also offers the advantage of controlled release of bioactive compounds. The encapsulated compounds can be designed to be released gradually, allowing for a sustained release of the bioactive compounds in the gastrointestinal tract or targeted delivery to specific sites in the body. This controlled release mechanism enhances the bioavailability and effectiveness of the bioactive compounds, ensuring that they exert their desired health benefits.

Furthermore, nanoencapsulation enables the incorporation of bioactive compounds into a wider range of food matrices without negatively impacting the sensory attributes of the final product. The encapsulation process can be tailored to suit the specific requirements of the bioactive compound and the desired food system, ensuring compatibility and minimizing any potential negative interactions. This flexibility in formulation allows for the development of functional foods that are both nutritious and palatable, promoting consumer acceptance and compliance with healthy dietary choices.

2.5.3 IMPROVING BIOAVAILABILITY AND HEALTH BENEFITS

Nanoencapsulation plays a crucial role in enhancing the bioavailability of bioactive compounds, overcoming their limitations in solubility, stability, and absorption. By reducing the particle size to the nanoscale and providing a protective barrier, nanoencapsulation improves the interaction of bioactive compounds with biological systems, enhances their enzymatic hydrolysis, protects them from degradation, and facilitates their absorption across biological barriers. The increased surface area of nanoscale particles allows for improved enzymatic hydrolysis in the gastrointestinal tract. Enzymes can more efficiently access and break down the encapsulated bioactive compounds, leading to enhanced release and absorption. This increased enzymatic hydrolysis ensures that a higher proportion of the bioactive compound is available for physiological action, maximizing its bioavailability. Nanoencapsulation also provides a physical barrier that protects bioactive compounds from enzymatic degradation during digestion. The encapsulation matrix shields the bioactive compound from enzymes in the gastrointestinal tract, preserving its integrity and bioactivity. This protection extends the bioactive compound's lifespan, ensuring that it remains available for absorption and utilization in the body.

Furthermore, the nanoscale size of encapsulated particles enables them to bypass physiological barriers in the gastrointestinal tract. The particles can pass through the mucus layer and tight junctions in the intestinal epithelium, facilitating their absorption into the bloodstream. This enhanced permeability allows for improved bioavailability of the encapsulated bioactive compound and increases its potential to exert physiological effects in target organs or tissues. Moreover, nanoencapsulation enables targeted delivery of bioactive compounds to specific sites in the body. Surface modifications of the nanoparticles or incorporation of targeting ligands can facilitate interactions with specific receptors or transporters, directing the encapsulated bioactive compound to the desired location. Targeted delivery ensures that the bioactive compound reaches its intended site of action, increasing its bioavailability and optimizing its therapeutic effects. The improved bioavailability resulting from nanoencapsulation has significant implications for enhancing the health benefits of bioactive compounds. Higher bioavailability means that a greater proportion of the ingested bioactive compound is absorbed and utilized by the body, leading to enhanced physiological effects. For instance, bioactive compounds with antioxidant properties, such as polyphenols, when encapsulated at the nanoscale, exhibit increased radical-scavenging activity, protecting against oxidative stress-related diseases.

Furthermore, nanoencapsulation of vitamins and minerals enhances their absorption and utilization, addressing nutrient deficiencies and improving overall nutritional status. This is particularly relevant in populations where nutrient deficiencies are prevalent, as it provides a means to increase the bioavailability of essential nutrients and improve health outcomes. In addition to improved bioavailability, nanoencapsulation allows for the controlled release of bioactive compounds. By modulating the properties of the encapsulation matrix, such as composition, structure, and thickness, the release rate of the bioactive compound can be tailored. Controlled release systems enable sustained or targeted delivery of the bioactive compound, ensuring a prolonged therapeutic effect and optimizing its utilization.

2.5.4 APPLICATIONS OF NANOENCAPSULATION IN NUTRITION

Nanoencapsulation has found numerous applications in the field of nutrition and functional foods. The ability to protect, stabilize, and improve the bioavailability of bioactive compounds has opened up new possibilities for incorporating them into a wide range of food products. The following are some examples of how nanoencapsulation is being applied:

- **Fortification of food products:**

  Nanoencapsulation offers a promising solution for fortifying food products with essential nutrients, vitamins, and minerals that are susceptible to degradation or have low solubility. By encapsulating these compounds at the nanoscale, their stability and solubility can be improved, allowing for their incorporation into a wide range of food matrices while maintaining the sensory attributes of the final product. Many essential nutrients, such as vitamins and minerals, are sensitive to environmental factors such as heat, light, oxygen, and pH. These factors can cause degradation, leading to a loss of nutritional value in food products. Nanoencapsulation provides a protective barrier around these compounds, shielding them from external factors and preserving their stability. The encapsulation matrix acts as a physical barrier, preventing direct contact between the encapsulated compounds and the surrounding environment, thus minimizing degradation during processing, storage, and consumption (Abdollahzadeh et al., 2022; Ahmad, 2019b; Fröhlich et al., 2019).

  Furthermore, the small particle size of nanoencapsulated compounds improves their solubility in aqueous and lipid-based food systems. Many essential nutrients, such as fat-soluble vitamins and certain minerals, have low water solubility, which limits their incorporation into water-based food products. Nanoencapsulation increases the surface area of the encapsulated compounds, facilitating their dispersibility and solubility in both aqueous and lipid-based food matrices. This enhanced solubility allows for better distribution and dispersion of the encapsulated compounds throughout the food product, ensuring their uniform presence and availability for absorption by the consumer. The ability to fortify food products with encapsulated essential nutrients, vitamins, and minerals opens up opportunities for the development of functional foods. Functional foods are designed to provide specific health benefits beyond basic nutrition. By incorporating encapsulated bioactive compounds into food products, manufacturers can create functional foods with enhanced nutritional profiles.

For example, calcium-fortified beverages, iron-fortified cereals, or omega-3 fatty acid-enriched snacks can be developed using nano-encapsulation techniques. These fortified products offer improved nutritional value and targeted health benefits, catering to consumer demands for healthier and more functional food choices.

Moreover, nanoencapsulation allows for the controlled release of encapsulated compounds, enabling sustained delivery of essential nutrients, vitamins, and minerals in the body. The encapsulation matrix can be designed to release the encapsulated compounds gradually over time, providing a sustained and prolonged release profile. This controlled release feature is particularly advantageous for certain nutrients that require sustained delivery to optimize their absorption and utilization in the body. The use of nanoencapsulation in fortifying food products offers numerous advantages. It provides a means to overcome the limitations of stability and solubility associated with essential nutrients, vitamins, and minerals. By encapsulating these compounds, their shelf life can be extended, and their bioavailability can be enhanced. Additionally, nanoencapsulation enables the development of functional foods with improved nutritional profiles, addressing specific health concerns and meeting the demands of health-conscious consumers. However, it is important to note that the safety and regulatory aspects of using nanoencapsulated compounds in food products need to be thoroughly evaluated. The potential for the release of nanoscale particles during digestion and their long-term effects on human health are areas of active research and concern. Regulatory authorities worldwide are monitoring the use of nanotechnology in food applications and establishing guidelines to ensure consumer safety.

- **Controlled release systems:**

Nanoencapsulation provides a versatile platform for the design of controlled release systems, enabling the sustained delivery and targeted release of bioactive compounds. This approach has significant implications for the development of functional foods and therapeutic applications. Controlled release systems allow for the gradual and sustained release of bioactive compounds over an extended period of time. By modulating the properties of the encapsulation matrix, such as its composition, structure, and thickness, the release rate of the encapsulated compound can be tailored to meet specific requirements. This sustained release profile ensures a continuous and prolonged presence of the bioactive compound, enhancing its therapeutic efficacy and providing long-term benefits. One notable application of controlled release systems is the incorporation of encapsulated probiotics into functional foods. Probiotics are live microorganisms that confer health benefits when consumed in adequate amounts (Alam et al., 2021). However, their viability and functionality can be compromised during processing and storage of food products due to factors such as exposure to heat, moisture, and oxygen. Nanoencapsulation offers a protective barrier for probiotics, shielding them from detrimental environmental conditions and enabling their survival until consumption. The encapsulated probiotics can be released gradually in the gastrointestinal tract, ensuring their delivery to the gut in a viable form. This controlled release system enhances the efficacy of probiotics in promoting gut health and provides a convenient means of incorporating these beneficial microorganisms into a wide range of food products.

Furthermore, controlled release systems can be employed to deliver bioactive compounds to specific organs or tissues in the body. By modifying the surface properties of the nanoparticles or incorporating targeting ligands, the encapsulated compounds can bc designed to interact with specific receptors or transporters, facilitating their uptake by target cells or tissues. This targeted release allows for localized delivery of the bioactive compound, maximizing its therapeutic effects while minimizing potential side effects in other areas of the body. For instance, nanoparticles encapsulating anticancer drugs can be designed to selectively accumulate in tumor tissues, delivering the therapeutic agent directly to the site of action and minimizing damage to healthy cells. Controlled release systems can also be utilized to overcome the challenges associated with the poor bioavailability and rapid clearance of certain bioactive compounds. By encapsulating these compounds, their release can be modulated to achieve optimal absorption and utilization in the body. This is particularly relevant for hydrophobic compounds or compounds with limited solubility in aqueous environments. The encapsulation matrix provides a solubilizing environment for these compounds, allowing for their gradual release and improved bioavailability. This controlled release approach ensures that the bioactive compounds are available at the desired site of action for an extended period, enhancing their therapeutic or functional effects.

- **Flavor and aroma enhancement:**

  Nanoencapsulation offers a valuable approach to the encapsulation of volatile compounds responsible for flavors and aromas in food products. These volatile compounds, which contribute to the sensory attributes and overall quality of food, are often susceptible to degradation and loss during processing, storage, and distribution. By encapsulating these compounds, their stability can be enhanced, and their controlled release can be achieved, leading to improved sensory attributes and extended shelf life of food products. The encapsulation of volatile compounds involves the formation of a protective shell or matrix around the compound, which acts as a barrier against external factors that can cause degradation, such as oxygen, light, heat, and moisture. The encapsulation materials can be selected based on their compatibility with the volatile compound and the desired release profile. Common encapsulation materials used for volatile compounds include lipids, proteins, and polysaccharides. One key advantage of nanoencapsulation in preserving volatile compounds is the physical barrier provided by the encapsulation matrix. This barrier prevents direct contact between the volatile compound and the surrounding environment, minimizing the risk of oxidation, evaporation, or chemical reactions that can lead to the loss of volatile flavors and aromas. The encapsulated volatile compounds remain protected until released under specific conditions, ensuring their integrity and preserving the desired sensory characteristics of the food product.

Furthermore, nanoencapsulation enables a controlled release of the encapsulated volatile compounds, allowing for their gradual and controlled release during consumption or food preparation. The encapsulation matrix can be designed to release the volatile compounds in response to specific triggers, such as changes in temperature, pH, or mechanical forces. This controlled release mechanism ensures that the volatile compounds are released at the desired time and in the desired amount, enhancing the sensory experience for consumers. The use of nanoencapsulation for encapsulating volatile compounds offers several benefits for the food industry. Firstly, it allows for the development of food products with improved sensory attributes. The encapsulation of volatile flavors and aromas preserves their intensity, ensuring that the desired sensory characteristics are maintained throughout the shelf life of the product. This is particularly important for products with extended shelf life or those that undergo processing techniques that can affect the volatile compounds, such as high-temperature treatments or long storage periods.

Secondly, nanoencapsulation enhances the stability of volatile compounds during processing and storage. By protecting the volatile compounds from degradation, their quality and shelf life are extended, resulting in a more consistent and desirable sensory experience for consumers. This is particularly relevant for food products that require long shelf life or those that undergo distribution to distant markets. Moreover, the controlled release of encapsulated volatile compounds allows for customized sensory experiences. Food manufacturers can design products with specific release profiles, ensuring that the volatile flavors and aromas are released at the optimal time and intensity. This customization provides flexibility in product development, allowing for the creation of unique flavor profiles and meeting consumer preferences. Overall, nanoencapsulation provides a valuable solution for encapsulating volatile compounds responsible for flavors and aromas in food products. By protecting these compounds from degradation and enabling their controlled release, nanoencapsulation enhances the sensory attributes of food products and extends their shelf life. This technology opens up opportunities for the development of innovative food products with enhanced flavors, aromas, and overall sensory experiences, meeting consumer demands for high-quality and enjoyable food products.

- **Nutraceuticals and dietary supplements:**

  Nanoencapsulation has gained significant attention in the development of nutraceuticals and dietary supplements due to its ability to improve the delivery and efficacy of bioactive compounds. Nutraceuticals are products that combine the nutritional benefits of food with the potential health benefits of pharmaceuticals, while dietary supplements are products intended to supplement the diet with specific nutrients or bioactive compounds. By encapsulating bioactive compounds in nanoscale carriers, nanoencapsulation offers several advantages for the development of these products, including enhanced stability, solubility, bioavailability, and the ability to deliver multiple bioactive compounds simultaneously. One of the key benefits of nanoencapsulation in the development of nutraceuticals and dietary supplements is its ability to enhance the stability of bioactive compounds. Many bioactive compounds, such as vitamins, polyphenols, and omega-3 fatty acids, are prone to degradation when exposed to environmental factors, such as oxygen, light, heat, and moisture. By encapsulating these compounds in nanoscale carriers, their exposure to these factors can be minimized, preserving their integrity and bioactivity. The encapsulation matrix acts as a protective barrier, shielding the bioactive compounds from external influences and ensuring their stability during processing, storage, and consumption.

In addition to stability enhancement, nanoencapsulation improves the solubility of bioactive compounds. Many bioactive compounds have limited solubility in water or other food matrices, which can hinder their incorporation into products and their absorption in the body. By reducing the particle size to the nanoscale, the surface area-to-volume ratio of the encapsulated compounds increases, leading to improved dispersibility and solubility. This enhanced solubility enables better incorporation of the bioactive compounds into various food matrices, such as beverages, powders, or capsules, enhancing their bioavailability and ensuring their maximum health benefits for consumers. Bioavailability is a critical factor in the effectiveness of nutraceuticals and dietary supplements. It refers to the extent and rate at which a bioactive compound is absorbed and becomes available for physiological action in the body. Many bioactive compounds have low bioavailability due to factors such as poor solubility, instability, and limited absorption in the gastrointestinal tract. Nanoencapsulation has been shown to enhance the bioavailability of these compounds by addressing these limitations. The small particle size of nanoencapsulated bioactive compounds plays a crucial role in improving their bioavailability. The nanoscale particles have a larger surface area, facilitating better interaction with the digestive enzymes and enhancing their enzymatic hydrolysis. This results in increased release and absorption of the bioactive compounds in the gastrointestinal tract. The encapsulation matrix also protects the compounds from enzymatic degradation, allowing for their efficient absorption and utilization in the body. Furthermore, nanoencapsulation enables the delivery of multiple bioactive compounds simultaneously, which can lead to synergistic effects and personalized nutrition. Nutraceuticals and dietary supplements often contain a combination of bioactive compounds that work together to produce beneficial health effects. Nanoencapsulation allows for the encapsulation of multiple compounds within the same nanocarrier, enabling their simultaneous delivery to the target site. This co-delivery of bioactive compounds can enhance their interactions, leading to synergistic effects and improved efficacy. Moreover, the encapsulation of multiple compounds in a single nanocarrier facilitates personalized nutrition by allowing for the customization of nutrient combinations based on individual needs and health goals.

- **Functional food packaging:**

  Nanoencapsulation offers promising opportunities for improving food packaging materials and enhancing the preservation and quality of food products. Packaging plays a crucial role in protecting food from physical, chemical, and microbial degradation, thereby extending its shelf life and maintaining its quality. By incorporating encapsulated bioactive compounds into packaging films, nanoencapsulation can provide active functionality to packaging materials, offering advantages over traditional passive packaging systems. One of the key applications of nanoencapsulation in food packaging is the incorporation of antimicrobial agents. Encapsulating antimicrobial agents at the nanoscale allows for their controlled release and targeted delivery to the food surface. These encapsulated antimicrobial agents can effectively inhibit the growth of spoilage-causing microorganisms, such as bacteria, yeasts, and molds, thereby extending the shelf life of perishable foods. The encapsulation matrix acts as a protective barrier, preventing the premature release of antimicrobial agents and ensuring their sustained activity over time. This active packaging approach reduces the need for additional preservatives and can help maintain the safety and quality of food products.

In addition to antimicrobial agents, nanoencapsulation enables the incorporation of antioxidants into food packaging materials.

Antioxidants play a vital role in preventing oxidative degradation of food components, such as lipids, proteins, and vitamins, which can lead to rancidity, color changes, and loss of nutritional value. By encapsulating antioxidants within packaging films, their release can be controlled, and their efficacy can be maintained over an extended period. This active packaging approach helps to protect food products from oxidative deterioration, preserving their sensory attributes and nutritional quality. Moreover, nanoencapsulation can be utilized to incorporate oxygen scavengers into packaging films. Oxygen is a major contributor to food spoilage and degradation, as it promotes the growth of aerobic microorganisms and accelerates oxidative reactions. Encapsulated oxygen scavengers within packaging films can actively remove or reduce oxygen levels in the package, creating a modified atmosphere that inhibits microbial growth and slows down oxidative reactions. This active packaging approach can enhance the shelf life of oxygen-sensitive food products, such as meat, seafood, bakery goods, and beverages. The use of nanoencapsulation in food packaging materials also enables the controlled release of bioactive compounds directly to the food surface. The encapsulation matrix can be designed to respond to specific environmental conditions, such as temperature, pH, or moisture, triggering the release of encapsulated bioactive compounds when needed. For example, antimicrobial agents can be released in response to microbial contamination, ensuring their targeted action on the food surface. This controlled release mechanism enhances the effectiveness of bioactive compounds, optimizing their performance and minimizing their waste.

Furthermore, nanoencapsulation in food packaging can contribute to sustainability efforts by reducing the need for excessive additives and preservatives. By incorporating encapsulated bioactive compounds, such as antimicrobial agents and antioxidants, into packaging films, the reliance on synthetic additives and preservatives can be reduced. This approach aligns with the growing consumer demand for clean-label and natural food products.

- **Personalized nutrition:**

  Nanoencapsulation technology provides a promising avenue for personalized nutrition by enabling the encapsulation of different bioactive compounds separately and combining them at the point of consumption. This approach offers the potential to develop customized formulations that cater to individual needs, delivering specific combinations of bioactive compounds in optimized dosages. Personalized nutrition recognizes that individuals have unique nutritional requirements based on factors such as age, gender, genetics, lifestyle, and health conditions. By encapsulating bioactive compounds separately, nanoencapsulation allows for the independent encapsulation of multiple compounds without interaction or degradation. This provides flexibility in formulating personalized combinations of bioactive compounds, taking into account specific nutritional deficiencies, health goals, and preferences. The ability to combine encapsulated bioactive compounds at the point of consumption offers several advantages. Firstly, it allows for the precise control of the dosage and ratio of different compounds, ensuring the delivery of specific nutrients in optimized amounts. This tailored approach ensures that individuals receive the bioactive compounds they need without the risk of overconsumption or inadequate intake.

Secondly, the separate encapsulation of bioactive compounds ensures their stability and protection from interactions or degradation during storage and processing. Each compound is encapsulated in its own carrier, maintaining its integrity and preserving its bioactivity until the point of consumption. This is particularly important for compounds that are sensitive to environmental factors, such as light, oxygen, or heat, which may affect their efficacy. Furthermore, combining encapsulated bioactive compounds at the point of consumption allows for enhanced bioavailability and synergistic effects. Different compounds can interact synergistically, leading to improved absorption, metabolism, and overall health benefits. For example, combining encapsulated vitamins with minerals or antioxidants can enhance their bioavailability and antioxidant capacity. By delivering these combinations in optimized dosages, personalized nutrition can maximize the potential synergistic effects between bioactive compounds. Nanoencapsulation also enables the incorporation of bioactive compounds with different solubilities into a single formulation. Since bioactive compounds can vary in their solubility in water or lipids, encapsulating them separately allows for the development of personalized formulations that can be easily mixed with various food matrices or beverages. This versatility in formulation ensures that individuals can consume their personalized nutrition in a convenient and enjoyable manner.

Moreover, the encapsulation of bioactive compounds at the nanoscale offers the possibility of targeted delivery. By modifying the surface properties of the encapsulated carriers or incorporating targeting ligands, the bioactive compounds can be directed to specific sites in the body, such as the gastrointestinal tract or specific organs. This targeted delivery approach enhances the bioavailability and therapeutic efficacy of the encapsulated bioactive compounds, further optimizing their personalized nutritional benefits.

2.6 SAFETY AND REGULATORY CONSIDERATIONS

2.6.1 POTENTIAL RISKS AND SAFETY ASSESSMENT

Nanoencapsulation technology offers numerous benefits for improving the stability, solubility, and bioavailability of bioactive compounds. However, as with any emerging technology, it is essential to address the potential risks associated with nanoencapsulated food products and conduct comprehensive safety assessments to ensure consumer protection. One of the primary concerns related to nanoencapsulation is the potential for unintended biological effects. The small size and increased surface area of nanoparticles can result in altered physicochemical properties and interactions with biological systems. These changes may affect the toxicity, bioactivity, and behavior of encapsulated compounds. To evaluate the safety of nanoencapsulated food products, it is crucial to assess the potential hazards associated with the nanoparticles themselves and the encapsulated bioactive compounds. This assessment typically involves a multidisciplinary approach, combining toxicological, physicochemical, and exposure assessment studies.

Toxicological studies examine the effects of nanoparticles on various biological systems, including cell cultures, animals, and humans. These studies assess parameters such as cytotoxicity, genotoxicity, immunotoxicity, and organ-specific toxicity. Researchers investigate the potential for nanoparticles to induce inflammation, oxidative stress, and other adverse effects. Physicochemical characterization is another critical aspect of safety assessment. It involves analyzing the size, shape, surface charge, stability, and composition of nanoparticles. These characteristics can influence their behavior in biological systems and their potential interactions with cells, tissues, and organs. Exposure assessment is necessary to determine the likelihood and extent of human exposure to nanoparticles from nanoencapsulated food products. This assessment considers factors such as the intended use of the product, consumer habits, and potential release of nanoparticles during processing, storage, and disposal. By estimating exposure levels, it is possible to assess the potential risks associated with nanoencapsulated food products. For example, the European Food Safety Authority (EFSA) has established guidelines for the risk assessment of nanomaterials in food and feed. They emphasize the need for comprehensive safety assessments, considering the physicochemical properties of nanomaterials, their potential toxicity, exposure routes, and intended use. Similarly, the United States Food and Drug Administration (FDA) has issued guidance documents on the safety assessment of nanomaterials in food. The FDA recommends evaluating the potential toxicity of nanomaterials using appropriate toxicological studies and characterizing the physicochemical properties of nanomaterials to understand their behavior in biological systems. Several studies have also been conducted to assess the safety of nanoencapsulated food products. For example, a study published in the Journal of Agricultural and Food Chemistry in 2019 evaluated the potential cytotoxicity and genotoxicity of nanoencapsulated curcumin, a bioactive compound with antioxidant properties. The results indicated that the nanoencapsulation process did not significantly affect the safety profile of curcumin. Another study published in the journal Nanotoxicology in 2018 investigated the potential toxicity of silica nanoparticles used for encapsulating bioactive compounds. The study found that the nanoparticles did not induce significant adverse effects in rats after oral administration.

Risk assessment combines the findings from toxicological, physicochemical, and exposure assessment studies to evaluate the overall safety of nanoencapsulated food products. The goal is to identify potential hazards, estimate the risks associated with exposure to nanoparticles, and determine whether the benefits of using nanoencapsulation outweigh the potential risks. Risk management strategies can then be developed to minimize or mitigate any identified risks. It is worth noting that the safety assessment of nanoencapsulated food products is a dynamic process. As scientific knowledge advances and new evidence becomes available, ongoing monitoring and reassessment are necessary to ensure the continued safety of these products.

2.6.2 REGULATORY FRAMEWORK FOR NANOTECHNOLOGY IN FOOD

Regulatory frameworks play a crucial role in ensuring the safe and responsible development, production, and marketing of nanoencapsulated food products. Governments and regulatory agencies around the world are actively working to develop specific guidelines and regulations to address the unique challenges posed by nanotechnology in the food sector. The regulatory landscape for nanotechnology in food varies across different countries and regions. Some countries have specific regulations for nanotechnology, while others integrate nanomaterials into existing regulatory frameworks for food and food additives. At the international level, the Codex Alimentarius Commission, a joint body of the Food and Agriculture Organization (FAO) and the World Health Organization (WHO), is actively involved in addressing the safety and regulatory aspects of nanotechnology in food. The Codex is developing guidelines and standards to ensure the safety of nanomaterials used in food and the accurate labeling of nanoencapsulated food products. Several countries, such as the United States, European Union member states, Canada, and Australia, have also taken steps to regulate nanotechnology in food. These regulations often focus on safety assessments, labeling requirements, and risk management strategies.

In the United States, the Food and Drug Administration (FDA) has issued guidance documents for industry regarding the use of nanotechnology in food and food contact substances. The FDA expects manufacturers to conduct appropriate safety assessments for nanoencapsulated food products and has emphasized the importance of accurately labeling these products. In the European Union, nanoencapsulated food products fall under the scope of the novel food regulation. This regulation requires food businesses to seek pre-market approval for novel foods, including those that contain nanomaterials. The European Food Safety Authority (EFSA) provides guidance and assesses the safety of nanoencapsulated food products before they can be authorized for sale in the EU market. Canada has also developed guidelines for the safety assessment and regulation of nanomaterials in food. Health Canada evaluates the potential risks associated with nanoencapsulated food products and provides guidance to manufacturers on safety assessment and labeling requirements. Australia has adopted a similar approach, requiring manufacturers to seek pre-market approval for novel foods, including those containing nanomaterials. The Australian regulatory framework focuses on safety assessments and the accurate labeling of nanoencapsulated food products. In addition to these examples, many other countries have initiated efforts to regulate nanotechnology in food. However, it is important to note that regulatory frameworks for nanoencapsulated food products are still evolving, and there is ongoing discussion and debate on the most appropriate approaches to ensure safety and consumer confidence. Collaboration among regulatory agencies, scientific communities, and industry stakeholders is vital for the effective regulation of nanoencapsulation in the food sector. This collaboration facilitates the exchange of information, the harmonization of regulatory approaches, and the development of international standards and guidelines.

2.7 FUTURE DIRECTIONS AND CHALLENGES

2.7.1 EMERGING TRENDS IN NANOTECHNOLOGY FOR THE FOOD INDUSTRY

Nanotechnology has the potential to revolutionize the food industry by offering innovative solutions for various challenges related to food production, processing, safety, and nutrition. As the field of nanotechnology continues to advance, several emerging trends are shaping its future applications in the food industry.

2.7.1.1 NANOSENSORS AND FOOD SAFETY MONITORING

One emerging trend is the development of nanosensors for food safety monitoring. Nanosensors can detect and quantify contaminants, pathogens, and spoilage indicators at low concentrations, providing rapid and sensitive detection methods. These nanosensors can be incorporated into smart packaging materials or used as handheld devices for on-site testing, enabling real-time monitoring of food quality and safety. For example, nanosensors based on carbon nanotubes or nanoscale metal oxides have been developed to detect various foodborne pathogens, such as *Salmonella* and *E. coli*. These nanosensors offer advantages such as high sensitivity, specificity, and rapid response time, improving the efficiency of food safety testing.

2.7.1.2 INTELLIGENT PACKAGING SYSTEMS

Intelligent packaging systems incorporating nanotechnology are another emerging trend in the food industry. These systems use nanomaterials and nanosensors to monitor and maintain the quality of packaged food products. Nanomaterials can be incorporated into packaging films to provide barrier properties, such as oxygen and moisture resistance, extending the shelf life of perishable foods. Additionally, nanosensors integrated into packaging materials can detect changes in temperature, pH, and gas composition, providing real-time information about the condition of the packaged food. This enables timely decision-making regarding product freshness and safety.

2.7.1.3 NANODELIVERY SYSTEMS FOR NUTRACEUTICALS

Nanotechnology offers exciting opportunities for the development of nanodelivery systems that improve the bioavailability and targeted delivery of nutraceuticals. Nanoencapsulation, as discussed earlier, is a promising approach for enhancing the stability, solubility, and bioavailability of bioactive compounds. In the future, nanodelivery systems may be tailored to specific consumer needs, enabling personalized nutrition. For instance, nanocarriers could be designed to encapsulate a combination of bioactive compounds, vitamins, and minerals, delivering them in optimized dosages based on individual requirements.

2.7.1.4 ENHANCED FUNCTIONAL PROPERTIES

Nanotechnology can also enhance the functional properties of food ingredients. For example, nanoemulsions and nanogels can improve the texture, stability, and sensory attributes of food products. Nanoscale particles can act as carriers for flavor and aroma compounds, providing controlled release and long-lasting sensory experiences. Nanotechnology can also be used to modify the structure and functionality of proteins and carbohydrates, enhancing their emulsifying, gelling, and foaming properties. This opens up new possibilities for creating novel food textures and structures.

2.7.2 ADDRESSING CHALLENGES AND ETHICAL CONCERNS

While the potential benefits of nanotechnology in the food industry are significant, there are also challenges and ethical concerns that need to be addressed for its responsible and sustainable implementation.

2.7.2.1 SAFETY ASSESSMENT AND REGULATION

One of the primary challenges is ensuring the safety of nanomaterials and nanoencapsulated food products. Comprehensive safety assessments, as discussed earlier, are essential to understand the potential risks and benefits associated with nanotechnology. Regulatory frameworks need to be established or updated to address the unique characteristics and challenges of nanomaterials in food. Regulatory authorities must define appropriate safety standards, labeling requirements, and guidelines for the use of nanotechnology in food products. Collaboration between industry, academia, and regulatory bodies is crucial to foster transparent communication and exchange of information regarding the safety and regulatory aspects of nanotechnology.

2.7.2.2 CONSUMER PERCEPTION AND ACCEPTANCE

Another challenge is consumer perception and acceptance of nanotechnology in food. There is a need for transparent communication and education about the benefits and risks of nanotechnology to build consumer trust. Consumer attitudes, beliefs, and concerns should be taken into account during the development and marketing of nanoencapsulated food products. Addressing ethical concerns related to nanotechnology is also important. Ethical considerations include ensuring equitable access to nanotechnology-enabled food products, avoiding potential environmental impacts, and respecting privacy rights when using nanosensors and intelligent packaging systems.

2.7.2.3 ENVIRONMENTAL IMPACT

The potential environmental impact of nanomaterials used in the food industry is an area of concern. The life cycle of nanomaterials, from production to disposal, needs to be carefully evaluated to minimize any adverse environmental effects. Efforts should be made to develop sustainable nanomaterial synthesis methods, reduce energy consumption, and consider the end-of-life management of nanomaterial-containing products. Additionally, the release of nanomaterials into the environment during food processing, packaging, and disposal should be monitored and controlled to prevent unintended consequences.

2.7.2.4 INTELLECTUAL PROPERTY AND ECONOMIC CONSIDERATIONS

Nanotechnology innovations in the food industry raise intellectual property and economic considerations. Protecting intellectual property rights encourages investment in research and development. However, striking a balance between protecting intellectual property and promoting collaboration and knowledge sharing is essential for the advancement of the field. Economic considerations also come into play, as the implementation of nanotechnology may require additional investments in research, infrastructure, and manufacturing capabilities. Identifying viable business models and ensuring fair economic benefits for all stakeholders are important aspects to consider. In conclusion, the future of nanotechnology in the food industry holds immense potential for improving food safety, nutrition, and sustainability. Emerging trends such as nanosensors for food safety monitoring, intelligent packaging systems, nanodelivery systems for nutraceuticals, and enhanced functional properties demonstrate the diverse applications of nanotechnology in food. Addressing challenges related to safety assessment, regulation, consumer perception, environmental impact, and ethical concerns is crucial for responsible and sustainable implementation. Collaboration between industry, academia, and regulatory bodies is essential to ensure the safe and ethical development, regulation, and adoption of nanotechnology in the food industry. By overcoming these challenges and leveraging the benefits of nanotechnology, the food industry can advance toward a future of enhanced food quality, safety, and consumer satisfaction.

2.8 CONCLUSION

In conclusion, the advancement of nanotechnology in the food industry offers numerous opportunities to address various challenges and improve various aspects of food production, processing, safety, and nutrition. Nanoencapsulation, nanosensors, intelligent packaging systems, and nanodelivery systems for nutraceuticals are just a few examples of the emerging trends that are shaping the future of nanotechnology in the food industry. Through nanoencapsulation, bioactive compounds can be protected, enhancing their stability, solubility, and bioavailability. This technology opens up possibilities for fortifying food products with essential nutrients, vitamins, and minerals, as well as developing customized formulations for personalized nutrition. Nanoencapsulation also allows for the controlled release of bioactive compounds, enabling sustained delivery and targeted release, thereby maximizing their therapeutic efficacy.

Nanosensors integrated into packaging materials and smart packaging systems enable real-time monitoring of food quality and safety. These technologies provide rapid and sensitive detection of contaminants, pathogens, and spoilage indicators, facilitating early intervention and reducing the risk of foodborne illnesses. Intelligent packaging systems incorporating nanotechnology can extend the shelf life of perishable foods by providing barrier properties and monitoring changes in temperature, pH, and gas composition. This helps maintain the quality and freshness of packaged food, reducing food waste and improving consumer satisfaction. Nanotechnology also contributes to the enhancement of functional properties in food products. Nanoemulsions, nanogels, and nanocarriers for flavor and aroma compounds improve the texture, stability, and sensory attributes of food. Modifications of proteins and carbohydrates at the nanoscale level enable the creation of novel food textures and structures.

While nanotechnology presents exciting opportunities for the food industry, it is important to address the challenges and ethical concerns associated with its implementation. Safety assessments, regulation, and transparent communication are essential to ensure the responsible use of nanomaterials and nanoencapsulated food products. Consumer perception, environmental impact, intellectual property rights, and economic considerations should also be taken into account to foster acceptance and sustainable development. Overall, the future of nanotechnology in the food industry holds great promise. By leveraging the benefits of nanotechnology and addressing the challenges, the food industry can enhance food safety, extend shelf life, improve nutrition, and create innovative and sustainable food products. Continued research, collaboration, and responsible implementation will play a crucial role in harnessing the full potential of nanotechnology to meet the evolving needs and expectations of consumers in the global food market.

KEYWORDS

- **food industry**
- **food packaging**
- **nanoencapsulation**
- **nanomaterials**
- **nanosensors**
- **nanotechnology**

REFERENCES

Abbas, K. A., Saleh, A. M., Mohamed, A., & MohdAzhan, N. (2009). The recent advances in the nanotechnology and its applications in food processing: A review. *Journal of Food, Agriculture and Environment, 7*, 14–17.

Abdollahzadeh, H., Fazlzadeh, M., Afshin, S., Arfaeinia, H., Feizizadeh, A., Poureshgh, Y., & Rashtbari, Y. (2022). Efficiency of activated carbon prepared from scrap tires magnetized by Fe_3O_4 nanoparticles: Characterisation and its application for removal of reactive blue19 from aquatic solutions. *International Journal of Environmental Analytical Chemistry, 102*, 1911–1925. https://doi.org/10.1080/03067319.2020.1745199.

Abdullahi, N., & Nura, A. (2018). Advances in food packaging technology—A review. *Journal Postharvest Technology 06*, 55–64. http://www.jpht.info.

Ahmad, K. S. (2019a). Arachis hypogaea activated carbon-assisted removal of 1-(4,6-dimethoxypyrimidin-2-yl)-3-(3-ethylsulfonylpyridin-2-yl)sulfonylurea herbicide in agriculturally adsorbed soils. *International Journal of Environmental Science and Technology, 16*, 6247–6258. https://doi.org/10.1007/s13762-018-2016-0.

Ahmad, K. S. (2019b). Arachis hypogaea derived activated carbon steered remediation of Benzimidazole based fungicide adsorbed soils. *Chemistry and Ecology, 35*, 576–591. https://doi.org/10.1080/02757540.2019.1600678.

Ahmad, K. S. (2020). Fungicidal methyl-2-benzimidazole carbamate adsorption in soil and remediation via prunus dulcis derived activated carbon. *Revista Internacional de Contaminación Ambiental, 36*, 429–442. https://doi.org/10.20937/RICA.53544.

Akgönüllü, S., Yavuz, H., & Denizli, A. (2021). Development of gold nanoparticles decorated molecularly imprinted–based plasmonic sensor for the detection of aflatoxin M1 in milk samples. *Chemosensors, 9*, 1–15. https://doi.org/10.3390/chemosensors9120363.

Alam, S., Khan, M. S., Umar, A., Khattak, R., Rahman, N. U., Zekker, I., Burlakovs, J., Dc Rubin, S. S., Ghangrekar, M. M., Bhowmick, G. D., Kallistova, A., Pimenov, N., Khan, A., & Zahoor, M. (2021). Preparation of pd–ni nanoparticles supported on activated carbon for efficient removal of basic blue 3 from water. *Water (Switzerland), 13*, 1–19. https://doi.org/10.3390/w13091211.

Bajpai, V. K., Kamle, M., Shukla, S., Mahato, D. K., Chandra, P., Hwang, S. K., Kumar, P., Huh, Y. S., & Han, Y. K. (2018). Prospects of using nanotechnology for food preservation, safety, and security. *Journal of Food Drug Analysis, 26*, 1201–1214. https://doi.org/10.1016/j.jfda.2018.06.011.

Barhoum, A., García-Betancourt, M. L., Jeevanandam, J., Hussien, E. A., Mekkawy, S. A., Mostafa, M., Omran, M. M., Abdalla, M. S., & Bechelany, M. (2022). Review on natural, incidental, bioinspired, and engineered nanomaterials: history, definitions, classifications, synthesis, properties, market, toxicities, risks, and regulations. *Nanomaterials, 12*. https://doi.org/10.3390/nano12020177.

Bashir, S., Thakur, A., Lgaz, H., Chung, I.-M., & Kumar, A. (2019). Computational and experimental studies on Phenylephrine as anti-corrosion substance of mild steel in acidic medium. *Journal of Molecular Liquids, 293*, 111539. https://doi.org/10.1016/j.molliq.2019.111539.

Bayda, S., Adeel, M., Tuccinardi, T., Cordani, M., & Rizzolio, F. (2020). The history of nanoscience and nanotechnology: From chemical-physical applications to nanomedicine. *Molecules, 25*, 1–15. https://doi.org/10.3390/molecules25010112.

Berekaa, M. M. (2015). Review article nanotechnology in food industry. *International Journal of Current Microbiology Application of Sciences, 4*, 345–357. http://www.ijcmas.com.

Biswas, R., Alam, M., Sarkar, A., Haque, M. I., Hasan, M. M., & Hoque, M. (2022). Application of nanotechnology in food: processing, preservation, packaging and safety assessment. *Heliyon, 8*, e11795. https://doi.org/10.1016/j.heliyon.2022.e11795.

Bobrinetskiy, I., Radovic, M., Rizzotto, F., Vizzini, P., Jaric, S., Pavlovic, Z., Radonic, V., Nikolic, M. V., & Vidic, J. (2021). Advances in nanomaterials-based electrochemical biosensors for foodborne pathogen detection. *Nanomaterials, 11*, 1–26. https://doi.org/10.3390/nano11102700.

Chellaram, C., Murugaboopathi, G., John, A. A., Sivakumar, R., Ganesan, S., Krithika, S., & Priya, G. (2014). Significance of nanotechnology in food industry. *APCBEE Procedia, 8*, 109–113. https://doi.org/10.1016/j.apcbee.2014.03.010.

Chung, I. M., Rajakumar, G., Gomathi, T., Park, S. K., Kim, S. H., & Thiruvengadam, M. (2017). Nanotechnology for human food: Advances and perspective. *Frontiers in Life Sciences, 10*, 63–72. https://doi.org/10.1080/21553769.2017.1365775.

Cicek, S., & Nadaroglu, H. (2015). The use of nanotechnology in the agriculture. *Advances Nano Research, 3*, 207–223. https://doi.org/10.12989/anr.2015.3.4.207.

Dąbrowska, S., Chudoba, T., Wojnarowicz, J., & Łojkowski, W. (2018). Current Trends in the Development of Microwave Reactors for the Synthesis of Nanomaterials in Laboratories and Industries: A Review. *Crystals, 8*. https://doi.org/10.3390/cryst8100379.

Dasgupta, N., et al. (2015). Nanotechnology in agro-food: From field to plate. *Food Research International, 69*, 381–400.

Dhewa, T. (2015). Octa journal of environmental research nanotechnology applications in agriculture: An update. *Octa Journal Environmental Research, 3*, 204–211. http://www.sciencebeingjournal.com.

Dhonchak, C., & Agnihotri, N. (2023). Computational insights in the spectrophotometrically 4h-chromen-4-one complex using DFT method. *Biointerface Research in Applied Chemistry, 13*, 357.

Elemike, E. E., Uzoh, I. M., Onwudiwe, D. C., & Babalola, O. O. (2019). The role of nanotechnology in the fortification of plant nutrients and improvement of crop production. *Applied Science, 9*, 1–32. https://doi.org/10.3390/app9030499.

Fahmy, H. M., Salah Eldin, R. E., Abu Serea, E. S., Gomaa, N. M., AboElmagd, G. M., Salem, S. A., Elsayed, Z. A., Edrees, A., Shams-Eldin, E., & Shalan, A. E. (2020). Advances in nanotechnology and antibacterial properties of biodegradable food packaging materials. *RSC Advances, 10*, 20467–20484. https://doi.org/10.1039/d0ra02922j.

Farhang, B. (2009). *Nanotechnology and applications in food safety*, 1st edn. Elsevier Inc. https://doi.org/10.1016/B978-0-12-374124-0.00022-3.

Figueroa-Lopez, K. J., Torres-Giner, S., Angulo, I., Pardo-Figuerez, M., Escuin, J. M., Bourbon, A. I., Cabedo, L., Nevo, Y., Cerqueira, M. A., & Lagaron, J. M. (2020). Development of active barrier multilayer films based on electrospun antimicrobial hot-tack food waste derived poly(3-hydroxybutyrate-co-3-hydroxyvalerate) and cellulose nanocrystal interlayers. *Nanomaterials, 10*, 1–24. https://doi.org/10.3390/nano10122356.

Fotie, G., Limbo, S., & Piergiovanni, L. (2020). Manufacturing of food packaging based on nanocellulose: Current advances and challenges. *Nanomaterials, 10*, 1–26. https://doi.org/10.3390/nano10091726.

Fröhlich, A. C., Foletto, E. L., & Dotto, G. L. (2019). Preparation and characterization of $NiFe_2O_4$/activated carbon composite as potential magnetic adsorbent for removal of ibuprofen and ketoprofen pharmaceuticals from aqueous solutions. *Journal of Cleaner Production, 229*, 828–837. https://doi.org/10.1016/j.jclepro.2019.05.037.

Garde-Cerdán, T., Souza-da Costa, B., Rubio-Bretón, P., & Pérez-Álvarez, E. P. (2021). Nanotechnology: Recent advances in viticulture and enology. *Journal of the Science of Food and Agriculture, 101*, 6156–6166. https://doi.org/10.1002/jsfa.11406.

Georgin, J., Pinto, D., Franco, D. S. P., Schadeck Netto, M., Lazarotto, J. S., Allasia, D. G., Tassi, R., Silva, L. F. O., & Dotto, G. L. (2022). Improved adsorption of the toxic herbicide diuron using activated carbon obtained from residual cassava biomass (*Manihot esculenta*). *Molecules, 27*. https://doi.org/10.3390/molecules27217574.

Ghidan, A. Y., & Al Antary, T. M. (2020). *Applications of nanotechnology in agriculture*. https://doi.org/10.5772/intechopen.88390.

Gómez-Arribas, L. N., Benito-Peña, E., Hurtado-Sánchez, M. del C., & Moreno-Bondi, M. C. (2018). Biosensing based on nanoparticles for food allergens detection. *Sensors (Switzerland), 18*, 1–32. https://doi.org/10.3390/s18041087.

Gunasekaran, S. (2014). Nanotechnology for food. https://doi.org/10.1002/9781118846315.ch8.

Hannon, J. C., Kerry, J., Cruz-Romero, M., Morris, M., Cummins, E. (2015). Advances and challenges for the use of engineered nanoparticles in food contact materials. *Trends in Food Science and Technology, 43*, 43–62. https://doi.org/10.1016/j.tifs.2015.01.008.

Hano, C., & Abbasi, B. H. (2022). Plant-based green synthesis of nanoparticles: Production, characterization and applications. *Biomolecules, 12*, 1–9. https://doi.org/10.3390/biom12010031.

Harish, V., Tewari, D., Gaur, M., Yadav, A. B., Swaroop, S., Bechelany, M., & Barhoum, A. Review on nanoparticles and nanostructured materials: bioimaging, biosensing, drug delivery, tissue engineering, antimicrobial, and agro-food applications. *Nanomaterials, 12*. https://doi.org/10.3390/nano12030457.

He, X., & Hwang, H. M. (2016). Nanotechnology in food science: Functionality, applicability, and safety assessment. *Journal of Food and Drug Analysis, 24*, 671–681. https://doi.org/10.1016/j.jfda.2016.06.001.

Huang, Y., Mei, L., Chen, X., & Wang, Q. (2018). Recent developments in food packaging based on nanomaterials. *Nanomaterials, 8*, 1–29. https://doi.org/10.3390/nano8100830.

Kalpana Sastry, R., Anshul, S., & Rao, N. H. (2013). Nanotechnology in food processing sector—An assessment of emerging trends. *Journal of Food Science and Technology, 50*, 831–841. https://doi.org/10.1007/s13197-012-0873-y.

Kaur, H., Thakur, R. C., & Kumar, H. (2020). Effect of proteinogenic amino acids L-serine/L-threonine on volumetric and acoustic behavior of aqueous 1-butyl-3-propyl imidazolium bromide at T = (288.15, 298.15, 308.15, 318.15) K. *Journal of Chemical Thermodynamics, 150*, 106211. https://doi.org/10.1016/j.jct.2020.106211.

Kaur, H., Thakur, R. C., Pathania, V. S., & Sharma, S. (2022). Effect of choline-based ionic liquid (cholinium ethanoate) on volumetric and acoustic properties of aliphatic amino acids (glycine and L-alanine). *Journal of Molecular Liquids, 346*, 118247. https://doi.org/10.1016/j.molliq.2021.118247.

Kaya, S., Thakur, A., & Kumar, A. (2023). The role of *in silico*/DFT investigations in analyzing dye molecules for enhanced solar cell efficiency and reduced toxicity. *Journal of Molecular Graphics and Modeling, 124*, 108536. https://doi.org/10.1016/j.jmgm.2023.108536.

King, T., Osmond-McLeod, M. J., & Duffy, L. L. (2018). Nanotechnology in the food sector and potential applications for the poultry industry. *Trends in Food Science and Technology, 72*, 62–73. https://doi.org/10.1016/j.tifs.2017.11.015.

Kraśniewska, K., Galus, S., & Gniewosz, M. (2020). Biopolymers-based materials containing silver nanoparticles as active packaging for food applications—A review. *International Journal of Molecular Sciences, 21*. https://doi.org/10.3390/ijms21030698.

Kulkarni, M. B., Ayachit, N. H., & Aminabhavi, T. M. (2022). Recent advancements in nanobiosensors: current trends, challenges, applications, and future scope. *Biosensors, 12*. https://doi.org/10.3390/bios12100892.

Kumar, A., & Thakur, A. (2020). *Encapsulated nanoparticles in organic polymers for corrosion inhibition*. Elsevier Inc. https://doi.org/10.1016/b978-0-12-819359-4.00018-0.

Kus-liśkiewicz, M., Fickers, P., & Ben Tahar, I. (2021). Biocompatibility and cytotoxicity of gold nanoparticles: Recent advances in methodologies and regulations. *International Journal of Molecular Sciences, 22* (2021). https://doi.org/10.3390/ijms222010952.

Lammari, N., Louaer, O., Meniai, A. H., & Elaissari, A. (2020). Encapsulation of essential oils via nanoprecipitation process: Overview, progress, challenges and prospects. *Pharmaceutics, 12*, 1–21. https://doi.org/10.3390/pharmaceutics12050431.

Lemes, A. C., Sala, L., Ores, J. D. C., Braga, A. R. C., Egea, M. B., & Fernandes, K. F. (2016). A review of the latest advances in encrypted bioactive peptides from protein-richwaste. *International Journal of Molecular Sciences, 17*. https://doi.org/10.3390/ijms17060950.
Li, W., Zhang, C., Chi, H., Li, L., Lan, T., Han, P., Chen, H., & Qin, Y. (2017). Development of antimicrobial packaging film made from poly(lactic acid) incorporating titanium dioxide and silver nanoparticles. *Molecules, 22*. https://doi.org/10.3390/molecules22071170.
Liu, G., Li, C., Stewart, B. A., Liu, L., Zhang, M., Yang, M., & Lin, K. (2020). Enhanced thermal activation of peroxymonosulfate by activated carbon for efficient removal of perfluorooctanoic acid. *Chemical Engineering Journal, 399*, 125722. https://doi.org/10.1016/j.cej.2020.125722.
Maluin, F. N., Hussein, M. Z., & Idris, A. S. (2020). An overview of the oil palm industry: Challenges and some emerging opportunities for nanotechnology development. *Agronomy, 10*. https://doi.org/10.3390/agronomy10030356.
Maluin, F. N., Hussein, M. Z., Nik Ibrahim, N. N. L., Wayayok, A., & Hashim, N. (2021). Some emerging opportunities of nanotechnology development for soilless and microgreen farming. *Agronomy, 11*, 1–28. https://doi.org/10.3390/agronomy11061213.
Martirosyan, A., & Schneider, Y. J. (2014). Engineered nanomaterials in food: Implications for food safety and consumer health. *International Journal of Environmental Research and Public Health, 11*, 5720–5750. https://doi.org/10.3390/ijerph110605720.
Mlalila, N., Kadam, D. M., Swai, H., & Hilonga, A. (2016). Transformation of food packaging from passive to innovative via nanotechnology: Concepts and critiques. *Journal of Food Science and Technology, 53*, 3395–3407. https://doi.org/10.1007/s13197-016-2325-6.
Mohammad, Z. H., Ahmad, F., Ibrahim, S. A., & Zaidi, S. (2022). Application of nanotechnology in different aspects of the food industry. *Discover Food, 2*. https://doi.org/10.1007/s44187-022-00013-9.
Mura, S., Seddaiu, G., Bacchini, F., Roggero, P. P., & Greppi, G. F. (2013). Advances of nanotechnology in agro-environmental studies. *Italian Journal of Agronony, 8*, 127–140. https://doi.org/10.4081/ija.2013.e18.
Musial, J., Krakowiak, R., Mlynarczyk, D. T., Goslinski, T., & Stanisz, B. J. (2020). Titanium dioxide nanoparticles in food and personal care products—What do we know about their safety? *Nanomaterials, 10*, 1–23. https://doi.org/10.3390/nano10061110.
Naresh, V., & Lee, N. (2021). A review on biosensors and recent development of nanostructured materials-enabled biosensors. *Sensors (Switzerland), 21*, 1–35. https://doi.org/10.3390/s21041109.
Neethirajan, S., Ragavan, V., Weng, X., & Chand, R. (2018). Biosensors for sustainable food engineering: Challenges and perspectives. *Biosensors, 8*. https://doi.org/10.3390/bios8010023.
Niculescu, A. G., & Grumezescu, A. M. (2022). Applications of chitosan-alginate-based nanoparticles—An up-to-date review. *Nanomaterials, 12* (2022). https://doi.org/10.3390/nano12020186.
Nidhi, H. K. (2022). Advancement in field of energy storage applications by using ionic liquids. *Journal of Physics: Conference Series, 2267*. https://doi.org/10.1088/1742-6596/2267/1/012043.

Nurfatihah, Z., & Siddiquee, S. (2019). Nanotechnology: Recent trends in food safety, quality and market analysis. *Nanotechnology of Applied Energy, Drug Food,* 283–293. https://doi.org/10.1007/978-3-319-99602-8_14.

Pandey, B. K., Dhar, R., & Pandey, K. B. (2022). Application of nanotechnology in food science, processing and packaging. https://doi.org/10.1007/978-3-030-98820-3.

Panpatte, D. G., & Jhala, Y. K. (2019). Nanotechnology for agriculture: Crop production & protection. https://doi.org/10.1007/978-981-32-9374-8.

Parveen, G., Bashir, S., Thakur, A., Saha, S. K., Banerjee, P., & Kumar, A. (2020). Experimental and computational studies of imidazolium based ionic liquid 1-methyl- 3-propylimidazolium iodide on mild steel corrosion in acidic solution Experimental and computational studies of imidazolium based ionic liquid 1-methyl- 3-propylimidazolium, *Materials Research Express, 7*, 016510. https://doi.org/https://doi.org/10.1088/2053-1591/ab5c6a.

Phata, J. (2010). *Nanotechnology in Agricultural Diseases and, Nanotechnology, 2*, 83–92.

Primoži˘c, M., Knez, Ž., & Leitgeb, M. (Bio). (2021). Nanotechnology in food science—Food packaging. *Nanomaterials, 11*, 292.

Puscaselu, R. G., Lobiuc, A., Dimian, M., & Covasa, M. (2020). Alginate: From food industry to biomedical applications and management of metabolic disorders. *Polymers (Basel), 12*, 1–30. https://doi.org/10.3390/polym12102417.

Qu, J., Liu, Y., Cheng, L., Jiang, Z., Zhang, G., Deng, F., Wang, L., Han, W., & Zhang, Y. (2021). Green synthesis of hydrophilic activated carbon supported sulfide nZVI for enhanced Pb(II) scavenging from water: Characterization, kinetics, isotherms and mechanisms. *Journal of Hazardous Materials, 403*, 123607. https://doi.org/10.1016/j.jhazmat.2020.123607.

Raja, I. S., Vedhanayagam, M., Preeth, D. R., Kim, C., Lee, J. H., & Han, D. W. (2021). Development of two-dimensional nanomaterials based electrochemical biosensors on enhancing the analysis of food toxicants. *International Journal of Molecular Sciences,* 22. https://doi.org/10.3390/ijms22063277.

Rashidi, L., & Khosravi-Darani, K. (2011). The applications of nanotechnology in food industry, *Critical Reviews in Food Science and Nutrition, 51*, 723–730. https://doi.org/10.1080/10408391003785417.

Ravichandran, R. (2010). Nanotechnology applications in food and food processing: Innovative green approaches, opportunities and uncertainties for global market. *International Journal of Green Nanotechnology: Physics and Chemistry, 1*, P72–P96. https://doi.org/10.1080/19430871003684440.

Riungu, N. J., Hesampour, M., Pihlajamaki, A., Manttari, M., Home, P. G., & Ndegwa, G. M. (2012). Removal of pesticides from water by nanofiltration. *Journal of Engineering and Applied Computer Science, 1*, 50–60.

Roopa, H., Panghal, A., Kumari, A., Chhikara, N., Sehgal, E., Rawat, K. Active packaging in food industry. *Novel Technologies in Food Science, 8*, 375–404. https://doi.org/10.1002/9781119776376.ch10.

Sahoo, M., Vishwakarma, S., Panigrahi, C., & Kumar, J. (2021). Nanotechnology: Current applications and future scope in food. *Food Frontiers, 2*, 3–22. https://doi.org/10.1002/fft2.58.

Samal, D. (2017). Use of nanotechnology in food industry: A review. *International Journal of Agriculture Environment and Biotechnology, 2*, 2270–2278. https://doi.org/10.22161/ijeab/2.4.90.

Sani, M. A., Azizi-Lalabadi, M., Tavassoli, M., Mohammadi, K., & McClements, D. J. (2021). Recent advances in the development of smart and active biodegradable packaging materials. *Nanomaterials, 11*, 1–34. https://doi.org/10.3390/nano11051331.
Shafiq, M., Anjum, S., Hano, C., Anjum, I., & Abbasi, B. H. (2020). An overview of the applications of nanomaterials and nanodevices in the food industry. *Foods, 9*, 1–27. https://doi.org/10.3390/foods9020148.
Shah, B. R., & Mraz, J. (2020). Advances in nanotechnology for sustainable aquaculture and fisheries. *Reviews in Aquaculture, 12*, 925–942. https://doi.org/10.1111/raq.12356.
Shang, Y., Hasan, K., Ahammed, G. J., Li, M., & Yin, H. (2019). Shang-2019-applications-of-nanotechnology-in-p (1).pdf. *Molecules, 24*.
Sharma, D., Abhinay Thakur, M. K. S. et al. (2014). Synthesis, electrochemical, morphological, computational and corrosion inhibition studies of 3-(5-Naphthalen-2-yl-[1,3,4]oxadiazol-2-yl)-pyridine against mild steel in 1 M HCl. *Asian Journal of Chemistry, 35*, 1079–1088. https://doi.org/https://doi.org/10.14233/ajchem.2023.27711.
Siddiqui, S., & Alrumman, S. A. (2021). Influence of nanoparticles on food: An analytical assessment. *Journal of King Saud Univiversity—Sciences, 33*, 101530. https://doi.org/10.1016/j.jksus.2021.101530.
Singh, A., Sharma, A., Ahmed, A., Sundramoorthy, A. K., Furukawa, H., Arya, S., & Khosla, A. (2021). Recent advances in electrochemical biosensors: Applications, challenges, and future scope. *Biosensors, 11*, 1–31. https://doi.org/10.3390/bios11090336.
Singh, T., Shukla, S., Kumar, P., Wahla, V., & Bajpai, V. K. (2017). Application of nanotechnology in food science: Perception and overview. *Frontiers in Microbiology, 8*, 1–7. https://doi.org/10.3389/fmicb.2017.01501.
Sozer, N., & Kokini, J. L. (2009). Nanotechnology and its applications in the food sector. *Trends in Biotechnology, 27*, 82–89. https://doi.org/10.1016/j.tibtech.2008.10.010.
Sridhar, K., Inbaraj, B. S., & Chen, B. H. (2021). Recent advances on nanoparticle based strategies for improving carotenoid stability and biological activity. *Antioxidants, 10*. https://doi.org/10.3390/antiox10050713.
Subhan, M., Choudhury, K., & Neogi, N. (2021). Advances with molecular nanomaterials in industrial manufacturing applications. *Nanomanufacturing, 1*, 75–97. https://doi.org/10.3390/nanomanufacturing1020008.
Thakur, A., & Kumar, A. (2021). Sustainable inhibitors for corrosion mitigation in aggressive corrosive media: A comprehensive study. *Journal of Bio- and Tribo-Corrosion, 7*, 1–48. https://doi.org/10.1007/s40735-021-00501-y.
Thakur, A., & Kumar, A. (2022). Recent advances on rapid detection and remediation of environmental pollutants utilizing nanomaterials-based (bio)sensors. *Science of the Total Environment, 834*, 155219. https://doi.org/10.1016/j.scitotenv.2022.155219.
Thakur, A., & Kumar, A. (2023). Recent trends in nanostructured carbon-based electrochemical sensors for the detection and remediation of persistent toxic substances in real- time analysis, Mater. Res. Express. 10 (2023) 034001. https://doi.org/https://doi.org/10.1088/2053-1591/acbd1a.
Thakur, A., Kaya, S., & Kumar, A. (2021). Recent innovations in nano container-based self-healing coatings in the construction industry. *Current Nanoscience, 18*, 203–216. https://doi.org/10.2174/1573413717666210216120741.
Thakur, A., Kumar, A., Sharma, S., Ganjoo, R., & Assad, H. (2022a). Materials Today : Proceedings Computational and experimental studies on the efficiency of Sonchus

arvensis as green corrosion inhibitor for mild steel in 0.5 M HCl solution. *Material Today: Proceedings, 66*, 609–621. https://doi.org/10.1016/j.matpr.2022.06.479.

Thakur, A., Kaya, S., Abousalem, A. S., Sharma, S., Ganjoo, R., Assad, H., & Kumar, A. (2022b). Computational and experimental studies on the corrosion inhibition performance of an aerial extract of Cnicus Benedictus weed on the acidic corrosion of mild steel. *Process Safety and Environmental Protection, 161*, 801–818. https://doi.org/10.1016/j.psep.2022.03.082.

Thakur, A., Kumar, A., Kaya, S., Vo, D. V. N., & Sharma, A. (2022c). Suppressing inhibitory compounds by nanomaterials for highly efficient biofuel production: A review. *Fuel, 312*, 122934. https://doi.org/10.1016/j.fuel.2021.122934.

Thakur, A., Kumar, A., Kaya, S., Marzouki, R., Zhang, F., & Guo, L. (2022d). Recent advancements in surface modification, characterization and functionalization for enhancing the biocompatibility and corrosion resistance of biomedical implants. *Coatings, 12*, 1459. https://doi.org/https://doi.org/10.3390/coatings12101459.

Thakur, A., Kaya, S., Abousalem, A. S., & Kumar, A. (2022e). Experimental, DFT and MC simulation analysis of Vicia Sativa weed aerial extract as sustainable and eco-benign corrosion inhibitor for mild steel in acidic environment. *Sustainable Chemistry and Pharmacy, 29*, 100785. https://doi.org/10.1016/j.scp.2022.100785.

Thakur, A., Kumar, A., & Zhang, R. (2023a). Alcoholic Beverage Purification Applications of Activated Carbon. In C. Verma & M. A. Quraishi (Eds.), *Activated Carbon: Progress and Applications*. The Royal Society of Chemistry. https://doi.org/10.1039/BK9781839169861-00152.

Thakur, A., Savaş, K., & Kumar, A. (2023b). Recent trends in the characterization and application progress of nano-modified coatings in corrosion mitigation of metals and alloys. *Applied Sciences, 13*, 730. https://doi.org/10.3390/app13020730.

Thakur, R. C., Kaur, R., & Kaur, H. (2020). Transport properties of thiaminehydrochloridein binary aqueous mixtures of galactose. *European Journal of Molecular & Clinical Medicine, 7*, 2020.

Thiruvengadam, M., Rajakumar, G., & Chung, I. M. (2018). Nanotechnology: Current uses and future applications in the food industry. *3 Biotech, 8*, 1–13. https://doi.org/10.1007/s13205-018-1104-7.

Ummi, A. S., & Siddiquee, S. (2019). Nanotechnology applications in food: Opportunities and challenges in food industry. *Nanotechnology: Applications in Energy, Drug and Food,* 295–308. https://doi.org/10.1007/978-3-319-99602-8_15.

Verma, C., Thakur, A., Ganjoo, R., Sharma, S., & Assad, H. (2023). Coordination bonding and corrosion inhibition potential of nitrogen-rich heterocycles: Azoles and triazines as specific examples. *Coordination Chemistry Reviews, 488*, 215177. https://doi.org/10.1016/j.ccr.2023.215177.

Wahab, M., Zahoor, M., Muhammad Salman, S., Kamran, A. W., Naz, S., Burlakovs, J., Kallistova, A., Pimenov, N., & Zekker, I. (2021). Adsorption-membrane hybrid approach for the removal of azithromycin from water: An attempt to minimize drug resistance problem. *Water (Switzerland), 13*. https://doi.org/10.3390/w13141969.

Weiss, J., Takhistov, P., & McClements, D. J. (2006). Functional materials in food nanotechnology. *Journal of Food Sciences, 71*, 107–116. https://doi.org/10.1111/j.1750-3841.2006.00195.x.

Wesley, S. J., Raja, P., Raj, A. A., & Tiroutchelvamae, D. (2014). Review on—nanotechnology applications in food packaging and safety. *International Journal of Engineering Research, 3*, 645–651. https://doi.org/10.17950/ijer/v3s11/1105.

Wróblewska-Krepsztul, J., Rydzkowski, T., Michalska-Pożoga, I., & Thakur, V. K. (2019). Biopolymers for biomedical and pharmaceutical applications: Recent advances and overview of alginate electrospinning. *Nanomaterials, 9*, 7–9. https://doi.org/10.3390/nano9030404.

Yadav, R. K., Singh, N. B., Singh, A., Yadav, V., Bano, C., Khare, S., & Niharika. (2020). Expanding the horizons of nanotechnology in agriculture: recent advances, challenges and future perspectives. *Vegetos, 33*, 203–221. https://doi.org/10.1007/s42535-019-00090-9.

Yu, H., Park, J. Y., Kwon, C. W., Hong, S. C., Park, K. M., & Chang, P. S. (2018). An overview of nanotechnology in food science: Preparative methods, practical applications, and safety. *Journal of Chemistry, 2018*. https://doi.org/10.1155/2018/5427978.

Zhan, J., Li, Z., Yu, G., Pan, X., Wang, J., Zhu, W., Han, X., & Wang, Y. (2019). Enhanced treatment of pharmaceutical wastewater by combining three-dimensional electrochemical process with ozonation to *in situ* regenerate granular activated carbon particle electrodes. *Separation and Purification Technology, 208*, 12–18. https://doi.org/10.1016/j.seppur.2018.06.030.

CHAPTER 3

Nano-Delivery Food Systems

NIDHI[1], RAMESH CHAND THAKUR[2,*], ASHISH KUMAR[3], and PRAVEEN KUMAR SHARMA[1]

[1]*Department of Chemistry, Lovely Professional University, Phagwara, Punjab, India*

[2]*Department of Chemistry, Himachal Pradesh University, Shimla, Himachal Pradesh, India*

[3]*NCE, Bihar Engineering University, Department of Science and Technology, Government of Bihar, India*

*Corresponding author

ABSTRACT

The nano-delivery system offers many benefits depending on the compatibility of the nanoparticle qualities with the bioactive and the target application. Basically, nano-delivery regulates solubility, food bioactive stability, bioavailability as well as their controlled release. There are specifically two types of nano-delivery systems: (a) solid delivery system including liquid nanoparticles, nanocrystals, and polymeric nanoparticles, and (b) liquid delivery system including nanoliposomes, nano-polyersomes, and nano-emulsions. While talking about the benefits of nano-delivery in the food system, they have a very wide range of applications including encapsulation of bioactive compounds, a system for the antibacterial properties, and toxicity of the food system. For effective applications in food and agriculture, toxicological studies combining *in silico*, *in vitro*, and *in vivo* studies are required to determine the safety of nano-delivery systems.

Nanotechnological Advancements in the Food Industry.
Ishrat Majid, Bababode Adesegun Kehinde, and Sheikh Shafat Hussain (Eds.)

DOI: 10.1201/9781774919750-3

This review aims to explore the nano deliveries including the characteristics, production techniques, advantages, and applications, particularly in food systems.

3.1 INTRODUCTION

By adopting a multiscale approach and becoming acquainted with the impact of material structure at the micro- and nanoscales, a significant study field has lately been opened up. Nanotechnology is a rapidly developing field that encompasses the creation, processing, and application of nanomaterials (George et al., 2023; Porter and Charman, 2001). Nanomaterials are the 'discrete entity' which lies in the nanometer range, that is, from 1 to 10 nm (Bouwmeester et al., 2009). Nanomaterials have had a crucial impact on the development of human civilization and are already being manufactured in the fields of electronics, pharmaceuticals, energy storage devices, and now food industries (Chau et al., 2007; George et al., 2022). At the nanoscale, it entails generating and modifying organic and inorganic materials. This might result in far more specialized and efficient ways to manipulate food polymers and polymeric components to improve food quality and safety. Also, this technology is used to improve the nutritional value of the food. Nanotechnology tools are used in the whole food chain starting from the cultivation to the packaging of the grown food. To precisely target nutritional and health advantages to certain sites in the human body or cells, nanotechnology promises to give a way to modify and manipulate food items (Acosta, 2009; Borel and Sabliov, 2014; Rana et al., 2023). This will allow for more effective and efficient delivery of nutrients, proteins, and antioxidants. For the application of these nanomaterials in food and agriculture, different nano-delivery systems are being used which enhance the control of the release of food at the proper time and place and the food quality. Nano-delivery of food further leads to the enhancement of the bioavailability and biodistribution which is dependent on the uptake, stability, and solubility of the components (Ouwmeester, 2015).

Different types of nano-delivery systems exist which are based on lipids, polysaccharides, polymers, and proteins. Out of all these systems lipids, polysaccharides, and protein-based systems are suited for the application in food because the other, that is, polymer-based nano-delivery systems are widely used in pharmaceuticals and biomedicals which elaborate many synthetic polymers which are not suitable for food (Bashir et al., 2020; Yeşil, 2006). These systems are explained as shown in Figure 3.1.

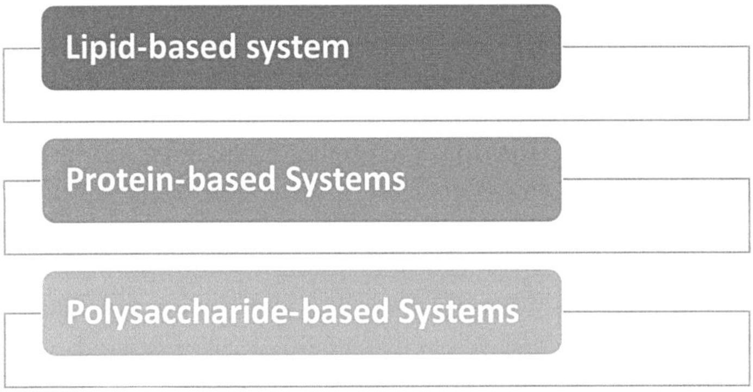

FIGURE 3.1 Chart showing different types of nano-delivery systems based on their composition.

1. **Lipid-based systems:**

   Lipid-based NDS is one of the most promising encapsulation strategies employed in the rapidly increasing field of nanotechnology. The ability to construct this nanoencapsulation system utilizing natural materials for industrial use, targetability, and the capacity to entrap compounds with various solubilities are just a few of its benefits. Nanoliposomes, nanocochelates, and archaesomes are the three primary lipid-based nanoencapsulation systems that can be employed to preserve and transport food and nutraceuticals (Kaur et al., 2017; Yeşil, 2006). Nano-delivery systems are mainly classified into two categories: liquid and solid nano-delivery systems which are further grouped according to their application and composition.

2. **Protein-based systems:**

   Food proteins are frequently utilized in food formulation because they are usually regarded as safe and have a high nutritional value. The most practical and often utilized matrix in culinary applications is protein hydrogel. Since proteins are proven to maintain strong nutritional qualities and a variety of capabilities, they have a particularly high potential as a delivery material. Whey proteins make an excellent delivery system for NDSs because of their many useful qualities, including their capacity to bind to hydrophobic bioactive compounds, ability to gel, and ability to emulsify. The bioavailability and chemical, physical, and biological stability of bioactive compounds may be improved by whey protein NDSs. These are particularly intriguing because they may form complexes with a range of nutrients that can be added, polysaccharides, lipids, or other biopolymers by different modifications (Patel et al., 2000).

3. **Polysaccharide-based systems:**

Natural polymers known as polysaccharides are derived from algae, plants, bacteria, some mammals, and microbes. They are composed of aldoses or ketoses linked together by glycosidic connections. Polysaccharides are commonly employed in foods and medications to create oral delivery systems for delicate bioactive substances because they are inexpensive, plentiful, safe, and useful. The properties and applications of nanoscale polysaccharide-based delivery vehicles, including their capacity to encapsulate, safeguard, and distribute bioactive substances, are discussed in this article. Systems of this kind are also referred to as poly (lactic) acid-based systems. Glycosidic linkages connect the monosaccharide monomers that make up polysaccharides. These polymers, which are molecules found naturally in plants, animals, and bacteria, are typically exceedingly large and branching. For the development of meals or nutraceuticals, targeted delivery systems and encapsulation based on polysaccharides are regarded to have promise. Due to their biodegradability and biocompatibility, homo- and copolymers made from poly(lactic acid) and poly(glycolic acid) are frequently utilized to create controlled release carriers for medicines and proteins (Nidhi, 2022; Parris et al., 2008). Improved water dispersibility, flavor masking, improved stability, lower volatility, and controlled release are just a few benefits of using nanoscale polysaccharide delivery systems; to deliver bioactive ingredients to specific intestinal sites with controlled release, polysaccharide nanocarriers can be built. Polysaccharides can also be combined with other biopolymers to create composite delivery systems with improved functional properties.

Nano-delivery systems are mainly classified into two categories: liquid and solid nano-delivery systems which are further grouped according to their application and composition (Figure 3.2).

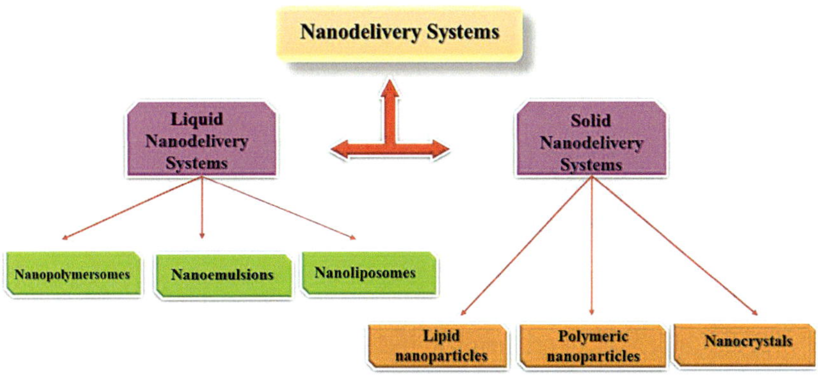

FIGURE 3.2 A flowchart showing various nano-delivery systems and their types.

1. **Liquid nano-delivery systems:**

   Liquid nano-delivery system is further classified into three subcategories:

   a. **Nano-polymersomes (NPS)**

   These are vesicles made of a layer of amphiphilic copolymers containing bioactive trapped in the cavity that has been created, much like liposomes. NPS are artificial vesicles with dimensions ranging from tens of nm to one m that include aqueous cavities generated by the self-assembly of amphiphilic copolymers. NPS can be tailored for use in a variety of biomedical applications, such as artificial organelles or drug delivery systems, because of its customizable characteristics. Several biodegradable and/or stimuli-responsive block copolymers are available, which are used to efficiently adjust the general characteristics of NPS, including drug encapsulation and release capacity. NPS are able to enclose hydrophobic and hydrophilic compounds in an aqueous core or a membrane bilayer, respectively, because of their adjustable characteristics. Their benefits over nano-lipid carriers include increased stability and adaptability as well as controlled release. NPS are viewed as possible attractive drug carriers in many clinical applications. Polymersomes are made from tri-block copolymers such as poly (caprolactone)-poly (ethylene glycol)-poly (caprolactone). Both hydrophilic and hydrophobic medicines can be captured with these. Enhanced stability and adaptability, controlled release, and polymer-carriers are its benefits (Kaur et al., 2020; Tanaka et al., 1988; Thakur and Kumar, 2021).

b. Nano-emulsions

Immiscible liquid mixes, such as water and oil, are examples of nano-emulsions. These nano-systems are often created using mechanical or chemical techniques. Chemical processes cause emulsion droplets to spontaneously develop as a result of the hydrophobic effects of lipophilic compounds in the presence of emulsifiers. In mechanical processes, high-energy approaches can be utilized to break up big emulsion droplets into smaller ones using a variety of mechanical events. The major distinction between conventional and nano-emulsions is the form and size of the particles dispersed in a suspension. Nano-emulsions typically have droplet diameters between 20 and 200 nm. Additionally, this is divided into emulsions and stabilized emulsions. An emulsion is the combination of two immiscible liquids, often water and oil. Surfactant-stabilized emulsions or other forms of stabilizing agents, such as polysaccharides, amphibian proteins, or phospholipids (Inoue et al., 2007). Nano-emulsions are highly used in food processing, to enhance the storage and stability of food products. These are also used as colorants for the enhancement of colors and quality of food products.

c. Nanoliposomes

Nano-liposomes are self-assembled, nanoscale vesicles made of one or more aqueous compartments enclosed in phospholipid bilayers. The droplet diameters of fine oil-in-water (o/w) dispersions range from 50 to 450 nm. Such nano-delivery devices could be created using an electrostatic deposition process that deposits layers at a time. This method involves incorporating charged polymers into a solution that already has a charged template structure. In addition to one or more lipid and phospholipid bilayers, the structure of liposomes may also include other molecules like proteins or polymers. Small unilamellar vesicles (SUVs) are liposomes with a single bilayer membrane that is up to 100 nm in size, while multilamellar large vesicles (MLVs) are liposomes with a membrane that is 500 nm to 5 um in size. The liposomes may concurrently absorb and release two different compounds with different solubilities (Inkyo et al., 2006; Kumar et al., 2020; Parmar and Thakur, 2006).

2. **Solid nano-delivery systems:**

Emulsions and solid nanoparticles are similar, however, the lipids in solid nanoparticles are in a different phase (Fee et al., 2003). These are physiological lipids that have been dispersed in water or aqueous surfactant solutions, resulting in sub-micron colloidal nanocarriers ranging in size from 50 to 1000 nm. Small size, vast surface area, excellent drug loading efficiency, and phase interaction at the interface are all benefits of such nanoparticles. The ability to entrap lipophilic molecules in stable particles without the need for organic solvents is one of the major benefits of solid lipid nanoparticles. Due to their distinct size-dependent characteristics and ability to integrate medicines, these nanoparticles provide a variety of therapeutic benefits (Ouwmeester, 2015; Patel, 2019). These benefits include the ability to produce on a large scale, the ability to combine hydrophilic and lipophilic pharmaceuticals, the high bioavailability of loaded medications, and low toxicity. Here, subcategories are further used to classify solid nano-delivery systems:

a. **Lipid nanoparticles:**

Although they exist in a solid state and are broken down more slowly than lipids in a liquid phase, solid lipid nanoparticles are comparable to emulsions. These particles have a poor loading capacity, and the medicine may be released during crystallization during storage (Cui et al., 2006).

The size of these delivery systems varies often based on the surfactant type and concentration; if one is too low or excessive, the system will combine. When using ionic surfactants, smaller particle sizes are typically achieved, and the stability of the emulsion is improved by using more than one surfactant (Sanguansri and Augustin, 2006). Lipid nanoparticles show a great application in food packaging in the food industry. They are used as a contaminant to oxidation and as an antimicrobial.

b. Polymeric nanoparticles

Polymeric nanoparticles are emulsifiers or surfactant-coated polymer matrices with trapped molecules. To increase the stability and bioavailability, natural polymer nanoparticles were created. Due to their wide surface area, high electrical conductivity, short path lengths, and high electrochemical activity for the transport of ions, these nanostructures are thought to perform better in energy storage than bulk conducting polymers (Bhat et al., 2020; Vaiserman et al., 2020). Polymeric nanoparticles exhibit good encapsulation efficiency as well as increased solubility and stability of hydrophobic medicines as a result of their physico-chemical properties. Due to these characteristics, the toxicity of loaded medications can be reduced, allowing for a controlled release at specific locations inside the body at relatively low doses. These are further divided into (a) nano-spheres and (b) nano-encapsulates. Polymer nano-capsules are nanoparticles in which the polymer encapsulates the bioactive by forming a wall around the core that is trapped within (Thakur and Sharma, 2017). Nano-spheres are made to prevent the bioactive from being degraded while it is confined within. as well as having increased intestinal permeability (Sakuma et al., 2002; Thakur et al., 2023). These help to increase the efficacy and water solubility of the food products.

c. Nanocrystals

A bioactive is enclosed in a surfactant in a nanocrystal. By raising the surface area to volume ratio, which accelerates the bioactive's *in vivo* dispersion, nanocrystals boost the solubility of medications that dissolve poorly in water. They enhanced the bioactive's cellular interaction and solubility in nanocrystal form (Aitken et al., 2007; Rashidi, 2021; Thakur et al., 2023). Nanocrystals are greatly used to enhance the shelf life of food and are also used in food packaging. Nanocrystals are pure (carrier-free) drug nanoparticle colloidal dispersion systems that are sub-micron in size (often between 10 and 800 nm). They can be made using mechanical or chemical processes. The main advantage of such nano-systems is that by lowering particle size to the nanoscale range, the particle surface area in contact with the dissolving liquid is increased. In comparison to pharmaceutical formulations that are currently in use, nano-crystal formulations are therefore thought to have potential therapeutic benefits. These advantages include higher drug loading as well as enhanced saturation solubility and dissolution rate (Jaiswal et al., 2015; Kakkar and Kaur, 2011).

3.2 APPLICATION IN FOOD INDUSTRY

In the food sector, nanotechnology presents appealing potential for improving food safety and quality control as well as for creating novel food additives, vitamins, and tastes. The food sector is changing thanks to nano-delivery. The majority of recorded uses of nano-delivery in food include:

1. Improvement of the food quality.
2. Alteration to the texture and structure of the food.
3. Detection of the micro-organisms in food products as well as used in food deterioration control.
4. Improved nutritious value of the food.
5. They are used as bioactive fortification.
6. They are also used as an antioxidant.
7. Shows a great application in food packaging.
8. Used in enhancement of the shelf life of the food as well as food protection.
9. Also used in detecting the neutralization of biochemical and chemical alteration.
10. Used as liquid carriers and enhances the solubility of the food products.

11. They are used as synthetic adhesives.
12. They are highly used in the removal of pathogens or contaminants.

3.3 CONCLUSION

The advancements in nanotechnology are currently receiving a lot of attention. In addition, as consumer demand for nutraceuticals, nutritional goods, and healthy food items grew throughout the years, the functional food industry grew as well. According to the study, it is recommended to assess the safety of nano-delivery systems for appropriate uses in the food sector and to determine the recommended daily consumption of nano-encapsulated nutraceuticals. Nano-delivery devices give a wide range of physical and chemical benefits for enhanced bioavailability and bioactivity. The biodegradability, biocompatibility, and nanoparticle ADME profile of the materials are used to assess the safety of nano-delivery systems. For the creation of economically feasible nano-delivery systems, further research needs to be done.

KEYWORDS

- **bioactive compounds**
- **food system**
- **nano-delivery**
- **nano-emulsions**
- **toxicity**

REFERENCES

Acosta, E. (2009). Bioavailability of nanoparticles in nutrient and nutraceutical delivery. *Current Opinion in Colloid & Interface Science, 14*, 3–15. https://doi.org/10.1016/j.cocis.2008.01.002

Aitken, R. J., Hankin, S. M., Tran, C. L., et al. (2007). REFNANO: Reference materials for engineered nanoparticle toxicology and metrology. Final report on Project CB01099.

Bashir, S., Thakur, A., Lgaz, H., et al. (2020). Corrosion inhibition efficiency of bronopol on aluminium in 0.5 M HCl solution: Insights from experimental and quantum chemical studies. *Surfaces and Interfaces, 20*, 100542. https://doi.org/10.1016/j.surfin.2020.100542

Bhat, I., Yathisha, U. G., Karunasagar, I., & Mamatha, B. S. (2020). Nutraceutical approach to enhance lutein bioavailability via nanodelivery systems. *Nutritions Review, 78*, 709–724. https://doi.org/10.1093/nutrit/nuz096

Borel, T., & Sabliov, C. M. (2014). Nanodelivery of bioactive components for food applications: Types of delivery systems, properties, and their effect on ADME profiles and toxicity of nanoparticles. *Annual Review of Food Science and Technology, 5*, 197–213. https://doi.org/10.1146/annurev-food-030713-092354

Bouwmeester, H., Dekkers, S., Noordam, M. Y., et al. (2009). Review of health safety aspects of nanotechnologies in food production. *Regulatory Toxicology and Pharmacology, 53*, 52–62. https://doi.org/10.1016/j.yrtph.2008.10.008

Chau, C. F., Wu, S. H., & Yen, G. C. (2007). The development of regulations for food nanotechnology. *Trends in Food Science and Technology, 18*, 269–280. https://doi.org/10.1016/j.tifs.2007.01.007

Cui, F., Shi, K., Zhang, L., et al. (2006). Biodegradable nanoparticles loaded with insulin-phospholipid complex for oral delivery: Preparation, *in vitro* characterization and *in vivo* evaluation. *Journal of Control Release, 114*, 242–250. https://doi.org/10.1016/j.jconrel.2006.05.013

Fee, M., Errington, N., Jumel, K., et al. (2003). Correlation of SEC/MALLS with ultracentrifuge and viscometric data for chitosans. *European Biophysics Journal, 32*, 457–464. https://doi.org/10.1007/s00249-003-0317-8

George, N., Singh, G., Singh, R., et al. (2022). Microwave accelerated green approach for tailored 1,2,3–triazoles via CuAAC. *Sustainable Chemistry and Pharmacy, 30*, 100824. https://doi.org/10.1016/j.scp.2022.100824

George, N., Singh, G., Singh, R., et al. (2023). Click modified bis-appended Schiff base 1,2,3-triazole chemosensor for detection of Pb(II)ion and computational studies. *Journal of Molecular Structure, 1288*, 135666. https://doi.org/10.1016/j.molstruc.2023.135666

Inkyo, M., Tahara, T., Iwaki, T., et al. (2006). Experimental investigation of nanoparticle dispersion by beads milling with centrifugal bead separation. *Journal of Colloid Interface and Science, 304*, 535–540. https://doi.org/10.1016/j.jcis.2006.09.021

Inoue, Y., Yoshimura, S., Tozuka, Y., et al. (2007). Application of ascorbic acid 2-glucoside as a solubilizing agent for clarithromycin: Solubilization and nanoparticle formation. *International Journal of Pharmaceutics, 331*, 38–45. https://doi.org/10.1016/j.ijpharm.2006.09.014

Jaiswal, M., Dudhe, R., & Sharma, P. K. (2015). Nanoemulsion: An advanced mode of drug delivery system. *3 Biotech, 5*, 123–127. https://doi.org/10.1007/s13205-014-0214-0

Kakkar, V., & Kaur, I. P. (2011). Evaluating potential of curcumin loaded solid lipid nanoparticles in aluminium induced behavioural, biochemical and histopathological alterations in mice brain. *Food and Chemical Toxicology, 49*, 2906–2913. https://doi.org/10.1016/j.fct.2011.08.006

Kaur, H., Thakur, R. C., & Kumar, H. (2020). Effect of proteinogenic amino acids L-serine/L-threonine on volumetric and acoustic behavior of aqueous 1-butyl-3-propyl imidazolium bromide at T = (288.15, 298.15, 308.15, 318.15) K. *Journal of Chemical Thermodynamics, 150*, 106211. https://doi.org/10.1016/j.jct.2020.106211

Kaur, K., Juglan, K. C., & Kumar, H. (2017). Investigation on Temperature-Dependent Volumetric and Acoustical Properties of Homologous Series of Glycols in Aqueous Sorbitol Solutions. *Journal of Chemical & Engineering Data, 62*, 3769–3782. https://doi.org/10.1021/acs.jced.7b00501

Kumar, H., Singh, G., Kataria, R., & Sharma, S. K. (2020). Volumetric, acoustic and infrared spectroscopic study of amino acids in aqueous solutions of pyrrolidinium based ionic liquid, 1–butyl–1–methyl pyrrolidinium bromide. *Journal of Molecular Liquids, 303*, 112592. https://doi.org/10.1016/j.molliq.2020.112592

Nidhi, Kaur H. (2022). Advancement in field of Energy storage applications by using Ionic Liquids. *Journal of Physics: Conference Series, 2267*. https://doi.org/10.1088/1742-6596/2267/1/012043

Ouwmeester, H. A. N. S. B. (2015). Review: A review of analytical methods for the identification and characterization of nano delivery systems in food, pp. 8231–8247.

Parmar, M. L., & Thakur, R. C. (2006). Effect of temperature on the partial molar volumes of some divalent transition metal sulphates and magnesium sulphate in the water-rich region of aqueous mixtures of ethylene glycol. *Journal of Molecular Liquids, 128*, 85–89. https://doi.org/10.1016/j.molliq.2005.12.011

Parris, N., Cooke, P. H., Moreau, R. A., & Hicks, K. B. (2008). Encapsulation of essential oils in zein nanospherical particles. *ACS Symposium Series 992*, 175–192. https://doi.org/10.1021/bk-2008-0992.ch010

Patel, G. B., Agnew, B. J., Deschatelets, L., et al. (2000). *In vitro* assessment of archaeosome stability for developing oral delivery systems. *International Journal of Pharmaceutics, 194*, 39–49. https://doi.org/10.1016/S0378-5173(99)00331-2

Patel, G. R. (2019). 済無No Title No Title No Title.

Porter, C. J. H., & Charman, W. N. (2001). *In vitro* assessment of oral lipid based formulations. *Advanced Drug Delivery Reviews, 50*, 127–147. https://doi.org/10.1016/S0169-409X(01)00182-X

Rana, S., Thakur, R. C., Kaur, H., et al. (2023). Investigating the solvation behavior of some lithium salts in binary aqueous mixtures of 1-ethyl-3-methylimidazolium tetrafluoroborate ([EMIM][BF4]) at equidistant temperatures (T = 298.15, 303.15, 308.15, 313.15, 318.15) K. *Journal of Chemical & Engineering Data*. https://doi.org/10.1021/acs.jced.3c00037

Rashidi, L. (2021). Different nano-delivery systems for delivery of nutraceuticals. *Food Bioscience, 43*, 101258. https://doi.org/10.1016/j.fbio.2021.101258

Sakuma, S., Suzuki, N., Sudo, R., et al. (2002). Optimized chemical structure of nanoparticles as carriers for oral delivery of salmon calcitonin. *International Journal of Pharmaceutics, 239*, 185–195. https://doi.org/10.1016/S0378-5173(02)00113-8

Sanguansri, P., & Augustin, M. A. (2006). Nanoscale materials development—A food industry perspective. *Trends in Food Science and Technology, 17*, 547–556. https://doi.org/10.1016/j.tifs.2006.04.010

Tanaka, K., Waki, H., Ido, Y., et al. (1988). Protein and polymer analyses up to m/z 100 000 by laser ionization time-of-flight mass spectrometry. *Rapid Communications in Mass Spectrometry, 2*, 151–153. https://doi.org/10.1002/rcm.1290020802

Thakur, A., & Kumar, A. (2021). *Sustainable inhibitors for corrosion mitigation in aggressive corrosive media: A comprehensive study*. Springer International Publishing.

Thakur, R. C., & Sharma, R. (2017). Viscometric studies of divalent transition metal sulphates in mixtures of water–diethylene glycol at 298.15–318.15 K. *Russian Journal of Physical Chemistry A, 91*, 1703–1709. https://doi.org/10.1134/S0036024417090254

Thakur, R. C., Sharma, A., Sharma, R., & Kaur, H. (2023). A comparative analysis of volumetric, viscometric and conductometric properties of triethylmethylammonium tetrafluoroborate (TEMABF4) and tetraethylammonium tetrafluoroborate (TEABF4) in pure propylene carbonate (PC) and binary aqueous propylene carbonate. *Journal of Molecular Liquids, 374*, 121244. https://doi.org/10.1016/j.molliq.2023.121244

Vaiserman, A., Koliada, A., Zayachkivska, A., & Lushchak, O. (2020). Nanodelivery of natural antioxidants: An anti-aging perspective. *Frontiers in Bioengineering and Biotechnology, 7*, 1–19. https://doi.org/10.3389/fbioe.2019.00447

Yeşil, N. (2006). Tiyatrodan ekranlara aktarilan gölge oyunu: Pepsi'nin reklam dizisinde "biz." *Milli Folk, 9*, 83–86. https://doi.org/10.1016/S1359-0294

CHAPTER 4

Food Nanosensors

VINAY KUMAR PANDEY[1,*], SHIVANGI SRIVASTAVA[2], RAFEEYA SHAMS[3], ANJALI TRIPATHI[4], and RAHUL SINGH[5]

[1]*Research and Development Cell, Biotechnology Department, Manav Rachna International Institute of Research and Studies (Deemed to Be University) Faridabad, Haryana, India*

[2]*Department of Food Technology, Harcourt Butler Technical University, Nawabganj, Kanpur, Uttar Pradesh, India*

[3]*Department of Food Technology and Nutrition, Lovely Professional University, Phagwara, Punjab, India*

[4]*School of Health Sciences and Technology, UPES University Dehradun, Dehradun, India*

[5]*Department of Bioengineering, Integral University, Lucknow, Uttar Pradesh, India*

*Corresponding author

ABSTRACT

Consumers, regulatory agencies, and the food business all have serious concerns about food safety and quality. A possible approach to overcoming these difficulties is the creation and use of food nanosensors. Foodborne pathogen detection, food quality parameter monitoring, and food contaminant identification are just a few of the current uses for food nanosensors, which span many facets of the food industry. Nanosensors have a number of benefits, including high sensitivity, quick detection, and the capacity to focus on certain analytes. They have been used to reduce the incidence of foodborne diseases by detecting bacterial infections like *Salmonella*, *Escherichia coli*,

Nanotechnological Advancements in the Food Industry.
Ishrat Majid, Bababode Adesegun Kehinde, and Sheikh Shafat Hussain (Eds.)

DOI: 10.1201/9781774919750-4

and *Listeria monocytogenes* and allowing for quick responses. As a way to supply safe and high-quality food products to consumers, nanosensors have also been used to monitor food quality indicators, such as pH, temperature, moisture, and freshness markers. By making it easier to identify and detect food pollutants including pesticides, heavy metals, and mycotoxins, nanosensors have improved food safety procedures. Surface-enhanced Raman scattering (SERS) nanosensors, optical, electrochemical, magnetic, and other forms of nanosensors are all used in the food sector. Real-time monitoring, remote sensing, and data-driven decision-making will be possible in the food business thanks to integration with Internet of Things (IoT) technology and data analytics platforms. An overview of the types, uses, and prospects for food nanosensors is given in this chapter.

4.1 INTRODUCTION

In today's world, when the demand for food is continually expanding and the global population is growing, food quality and safety are of the utmost significance. For the sake of sustaining the sustainability of food systems, customer confidence, and public health, it is essential to ensure the safety and quality of food (López et al., 2018). In this article, we will examine the history, importance, and factors affecting food safety and quality, as well as the steps used to accomplish them. The handling, preparation, and storage of food in a way that prevents the contamination and growth of hazardous microbes, chemicals, and other objects is referred to as "food safety." It includes a variety of behaviors, such as good hygiene, secure food handling, suitable cooking temperatures, and suitable storage settings. Foodborne infections can be caused by contaminated or unhealthy food, and they are a major global public health concern (Paul et al., 2018). Consuming food infected with harmful bacteria, viruses, parasites, or toxins results in foodborne illnesses, also known as food poisoning. Different symptoms, ranging from minor gastrointestinal discomfort to serious consequences and even death, might be brought on by these disorders (Bernardo et al., 2018). Millions of people throughout the world contract foodborne diseases each year, and the financial burden brought on by medical expenses, missed productivity, and food recalls is significant. Contrarily, food quality refers to the qualities that make a food palatable, desired, and nourishing. It includes a number of elements, such as sensory qualities (flavor, aroma, and texture), nutritional content, the lack of pollutants, and compliance with legal requirements. In addition to exceeding consumer expectations, high-quality food improves general health and well-being (Rahman et al., 2018).

Beyond just one person's health, food quality and safety are crucial. They have important effects on social welfare, economic growth, and public health. Building confidence among consumers, producers, and regulators depends on ensuring a safe and consistent food supply. It encourages the intake of a variety of healthy foods and increases customer confidence, supporting overall dietary patterns and reducing malnutrition (Wang et al., 2019). The necessity for strong food safety and quality systems has been further underscored in recent years by the globalization of food production and distribution. Food crosses international borders and goes through several supply chain stages, including manufacturing, processing, packaging, transportation, and retail. Each step presents the chance of contamination or a decline in quality. In order to protect public health and promote trade, international collaboration and harmonization of food safety standards and laws have become essential. Governments, regulatory bodies, and industry participants have put in place extensive systems and practices to guarantee food safety and quality. These include hazard analysis and critical control point (HACCP) systems for identifying and controlling risks at crucial stages of production, good manufacturing practices (GMPs) for food processing, and good agricultural practices (GAPs) for safe crop production. Furthermore, strict regulatory frameworks have been put in place to ensure compliance and hold individuals in charge of maintaining food safety and quality accountable. These frameworks include food safety legislation, labeling specifications, and traceability systems (Yen and Cheng, 2023). To monitor adherence to these standards and identify any variations or potential dangers, regular inspections, audits, and testing programs are carried out. Technology advancements have been extremely important in improving food quality and safety. The detection of diseases and pollutants can be done more quickly and accurately thanks to rapid diagnostic techniques, DNA sequencing, and real-time monitoring systems. In contrast to conventional heat-based approaches, novel preservation techniques include high-pressure processing, irradiation, and cold plasma treatment to preserve the nutritional value and sensory qualities of food while guaranteeing microbiological safety. The issues of guaranteeing food safety and quality persist despite tremendous improvements. New dangers are continuously being presented by emerging diseases, changing foodborne hazards, and the international scope of food supply chains. The environment of managing food safety and quality is further complicated by climate change, environmental degradation, and antibiotic resistance (Zhou et al., 2023).

A promising technology that has the potential to revolutionize several industries, including the food business, is nanosensors. These minuscule gadgets, whose sizes typically range from 1 to 100 nm, provide special skills for identifying and keeping track of particular substances or circumstances at the nanoscale. Nanosensors have the potential to improve food safety, quality control, and traceability in the context of the food industry, thereby answering the rising needs for safer and more environmentally friendly food production and consumption. In order to ensure the safety and quality of food items, the food business must overcome various obstacles. Pathogen contamination, chemical residue contamination, allergy contamination, freshness, spoilage, and adulteration problems are all serious difficulties (Chan et al., 2023). Traditional approaches to monitoring and detecting these pollutants frequently rely on subjective sensory evaluation, which may not be sensitive or specific enough, or on time-consuming laboratory analysis. With its ability to quickly, sensitively, and selectively detect pollutants and quality factors in real time, nanosensors represent a promising alternative. The detection of diseases that are found in food is one application where nanosensors have tremendous potential. Even minute quantities of bacteria, viruses, or other germs that can result in foodborne diseases can be found with these minuscule sensors. Nanosensors can target and precisely identify the presence of diseases by using particular recognition components, such as antibodies or DNA probes. This makes it possible to take preventive measures in a timely manner, lowering the likelihood of outbreaks and ensuring the quality of food items (Banerjee et al., 2023).

For the purpose of monitoring chemical contaminants in food, such as pesticides, heavy metals, and poisons, nanosensors can be quite useful. These pollutants have the potential to harm human health, thus it is important to properly check their presence in food. To quickly screen for and analyze these pollutants, nanosensors can be made to specifically detect and measure them. This enables quick action to stop the distribution of tainted products and lower the danger to the public's health by enabling early detection of contamination. Using nanosensors to ensure food quality and freshness can have a huge impact in another field. By detecting changes in factors like pH, temperature, humidity, and gas composition, nanotechnology-based sensors can provide real-time information on the quality and freshness of food products. The freshness of perishable commodities can be determined, for instance, by monitoring the amounts of oxygen, carbon dioxide, and volatile chemicals using nanosensors embedded in packing materials. This knowledge can be used to improve food quality overall, minimize food waste, and optimize storage conditions. Nanosensors have the potential to improve food supply chain traceability (Rutzky et al., 2019). Food packaging and labeling that include distinctive identifiers, such as quantum dots or nanobarcodes, make it simple to track and authenticate items along the entire supply chain. Improved traceability is made possible by this, making it possible to quickly determine a product's origin, spot fake items, and facilitate recalls when necessary. Improved traceability not only boosts consumer confidence but also makes it easier to spot and effectively resolve any possible safety or quality issues. Nanosensors can be used for targeted distribution of bioactive chemicals in food products in addition to their detection capabilities. Targeted distribution of functional elements, such as vitamins, antioxidants, or antimicrobial agents, can be accomplished by encapsulating bioactive molecules into nanoscale carriers, such as liposomes or nanoparticles. This can decrease the need for excessive amounts of additives or preservatives while improving the nutritional value, flavor, and shelf life of food products. Despite the tremendous promise of nanosensors in the food business, a number of issues must be resolved before they can be widely used. One major concern is the safety and potential risks associated with the use of nanomaterials in food. The impact of nanomaterials on human health and the environment requires thorough investigation and risk assessment (Hemdan et al., 2020).

The aim of this chapter is to give a general overview of how nanosensors are used in the food business and some examples of how they might be improved food safety, quality control, and traceability. The goal of this chapter is to examine the developments, difficulties, and potential applications of nanosensors in the food industry. The chapter's scope includes a variety of food nanosensor-related topics. It begins by explaining the idea of nanosensors and outlining the special qualities that make them appropriate for use in the food business. The next section of the chapter digs into the specific applications of nanosensors, including the detection of chemical pollutants, freshness indicators, and foodborne pathogens. Additionally, it covers how nanosensors might improve food product traceability and authenticity. The chapter also looks at how nanosensors may be used to deliver bioactive substances precisely where they are needed and how that might affect how well-balanced and long-lasting food products are nutritionally. It takes into account the difficulties and worries related to the use of nanosensors in the food sector, including safety issues, legal issues, and customer acceptance. To give readers a thorough grasp of the applications and prospective advantages of food nanosensors, the chapter integrates the latest research findings, case examples, and technological advancements in the industry. It intends to be a useful resource for academics, industry experts, and other interested parties in the food sector who wish to learn more about the possibilities provided by nanosensors for assuring the safety, quality, and traceability of food.

4.2 OVERVIEW OF NANOSENSORS

4.2.1 DEFINITION AND BASIC PRINCIPLES OF NANOSENSORS

Innovative tools called nanosensors make use of nanotechnology to identify and examine characteristics or substances at the nanoscale. They are made to engage with the target analyte or environment and generate quantifiable signals that can be decoded and analyzed. By identifying contaminants, keeping track of freshness, and evaluating the general quality of food products, nanosensors in the food sector are essential for maintaining food safety, quality control, and traceability. Nanosensors fundamental principles center on their capacity to recognize and react to changes in physical, chemical, or biological characteristics (Siddique and Shah, 2018). The sensor element, transducer, and readout system are typically their three basic parts. The target analyte, which could be a particular chemical, pathogen, or physical property like temperature or pH, interacts with the sensing element. The readout system analyzes and presents the data for analysis while the transducer transforms the sensing event into a quantifiable signal. The utilization of nanomaterials, such as nanoparticles or nanocomposites, which have unique features because of their small size and high surface-to-volume ratio, is one of the key ideas underlying nanosensors (Al-Ani et al., 2018). These nanomaterials can be designed to have particular properties that allow for quick and accurate detection. For instance, receptor molecules or probes that specifically attach to the target analyte can be added to nanoparticles to make them functional. This produces a signal that can be measured. The capacity of nanosensors to boost the signal or increase sensitivity through signal transduction pathways is another crucial concept. Several methods, including surface plasmon resonance, fluorescence resonance energy transfer, and electrochemical reactions, can be used to accomplish this. These mechanisms make nanosensors highly sensitive and able to deliver precise and dependable results since they enable the detection of even tiny quantities of analytes (Robles-López et al., 2018).

Nanosensors can also be made to be selective, allowing them to distinguish between several analytes or focus on certain molecules of interest. Utilizing specialized recognition elements with a high affinity and specificity for the target analyte, such as antibodies, enzymes, or aptamers, can help accomplish this selectivity. Nanosensors can interact with the target analyte, particularly by embedding these recognition features into the detecting element, reducing false positives or false negatives. In the food business, nanosensors provide a number of advantages over conventional sensing methods. They can be integrated into portable devices due to their tiny size, allowing for on-site and real-time monitoring. Additionally, they offer sensitive and speedy detection, facilitating quick judgement and cutting down on the time and resources needed for laboratory examination. Nanosensors can also save money because they frequently only need small amounts of samples and can be made utilizing scalable manufacturing techniques (Erdoğmus and Rowland, 2018).

4.2.2 TYPES OF NANOSENSORS USED IN FOOD ANALYSIS

4.2.2.1 OPTICAL NANOSENSORS

Optical nanosensors have drawn a lot of interest in the food business because of their distinctive qualities and capacity for detecting and analyzing a wide range of characteristics. These nanosensors monitor and quantify certain compounds or attributes in food products using optical signals, such as variations in light intensity, wavelength, or fluorescence. High sensitivity, selectivity, and real-time monitoring are just a few benefits that optical nanosensors provide, making them ideal tools for assuring food safety and quality control. A popular optical nanosensor based on surface plasmon resonance (SPR) is employed in the food sector. SPR happens when light strikes a metal surface and produces an evanescent wave that travels into the medium around it (Ward et al., 2018). The nanosensor can selectively attach to the target analyte, causing a change in the refractive index and resulting in a shift in the SPR signal by functionalizing the metal surface with certain receptors or probes, such as antibodies or aptamers. This shift can be quantified and associated with the analyte's quantity or presence, giving quantitative data. Another class of optical nanosensors that finds widespread use in the food business is fluorescent nanosensors. These nanosensors are made up of fluorescent molecules or quantum dots that, when stimulated by an external light source, emit light of a certain wavelength. The presence or interaction with the target analyte can modify the fluorescence's intensity or wavelength. The nanosensor can selectively bind to the analyte and cause a change in fluorescence, allowing for its detection and measurement, by integrating recognition components, such as molecular probes or enzymes. High sensitivity, quick reaction times, and the potential for multiplexed detection are all features of fluorescent nanosensors that make it possible to monitor several analytes at once (Xie et al., 2020). Other optical methods, like absorbance spectroscopy and Raman spectroscopy, have also been used in the food business in addition to SPR and fluorescence-based nanosensors. Nanomaterials that can absorb particular wavelengths of light are used in absorbance-based nanosensors, and the intensity of the absorbed light can be associated with the concentration of the target analyte. Raman spectroscopy includes measuring scattered light, which reveals details about the analyte's molecular structure and makeup. SERS nanosensors are nanoparticles with improved Raman scattering capabilities that can magnify the Raman signal for extremely sensitive analyte detection and identification (Chauhan et al., 2018) (Figure 4.1).

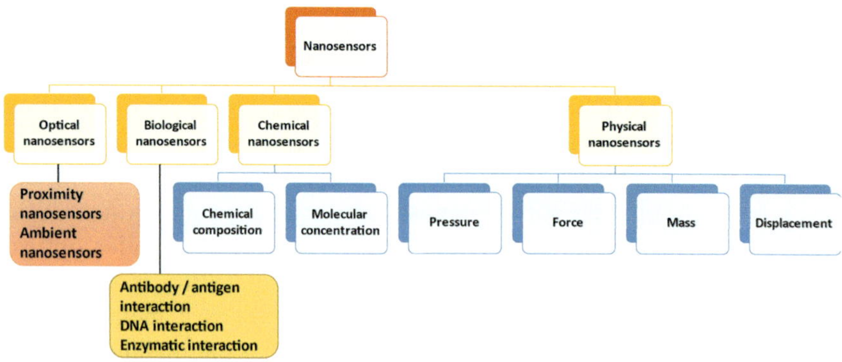

FIGURE 4.1 Types of nanosensors used in various industries.

Optical nanosensors have a wide range of uses in the food business. They can be used to ensure food safety and stop the ingestion of dangerous substances by identifying pollutants like diseases, poisons, or chemical residues. In order to determine freshness, spoilage, and storage conditions, optical nanosensors can also be used to detect food quality parameters like pH, temperature, humidity, and gas concentrations. They can also be used to examine nutritional content, such as determining the quantities of vitamins or nutrients, and to spot food fraud or adulteration. Despite optical nanosensors' many benefits, there are still some issues that need to be resolved. Assuring the stability and reproducibility of the nanosensors, defining standardized protocols for their integration into food industry processes, and building strong and trustworthy calibration methodologies are a few of these. Additionally, in order to guarantee customer safety, the possible toxicity and environmental impact of nanomaterials employed in nanosensors should be carefully assessed (Suárez et al., 2019).

4.2.2.2 ELECTROCHEMICAL NANOSENSORS

The potential of electrochemical nanosensors to detect and quantify particular analytes with great sensitivity and selectivity has attracted considerable interest in the food business. The target analyte interacts with the detecting surface in these nanosensors, causing changes in electrical impulses that are measured using electrochemical methods. Electrochemical nanosensors are useful tools for guaranteeing food safety and quality control because they have various benefits, such as quick response, miniaturization, and the ability for on-site and real-time monitoring (Yang, 2020). The amperometric sensor is a typical electrochemical nanosensor type. Using an electrochemical reaction between the target analyte and a sensing electrode, amplimetric nanosensors measure the current that results. Nanomaterials like nanoparticles, nanowires, or nanotubes are frequently used to modify sensing electrodes because they increase the surface area and make it easier to immobilize recognition components like enzymes or antibodies. An electrochemical reaction that results from the target analyte binding to the recognition element causes a change in current that may be measured and linked with the analyte concentration. The impedance sensor is another kind of electrochemical nanosensor frequently utilized in the food sector. Nanosensors that use electrical impedance to collect data track changes in the electrical impedance of the sensing interface as a result of interactions between the sensing layer and the target analyte. Nanomaterials with a high surface area for analyte binding, such as graphene, carbon nanotubes, or conductive polymers, are frequently used in the sensor layer. Target analyte binding modifies the electrical characteristics of the sensing layer, causing changes in impedance that can be identified and measured (Zhao et al., 2018).

In the food business, potentiometric nanosensors are also used to measure electrical potential changes brought on by interactions between a target analyte and a reference electrode. The ion-selective electrodes (ISEs) used in these nanosensors often bind to particular ions present in the food sample. It is possible to detect and correlate the potential change brought on by the binding of ions to the ISE with the ion concentration in the sample. In the food sector, electrochemical nanosensors have a wide range of uses. They are used to ensure food safety and quality control by detecting pollutants such as heavy metals, pesticides, and mycotoxins. Utilizing electrochemical nanosensors to measure variables like pH, temperature, and gas concentrations allows for the monitoring of food freshness, spoilage, and storage conditions (Li and Pasinetti, 2020). They are also employed for evaluating the authenticity and quality of food products, including the detection of adulterants, food fraud, and nutritional content. Electrochemical nanosensors have benefits, however, there are several issues that need to be resolved. These include developing standardized methods for their integration into food industry operations, enhancing the sensitivity and selectivity of the nanosensors, and ensuring their stability and reproducibility. Additionally, the potential toxicity and environmental impact of nanomaterials used in the nanosensors need to be carefully evaluated to ensure consumer safety (Park, 2018).

4.2.2.3 MAGNETIC NANOSENSORS

Due to their capacity to accurately measure and detect a variety of analytes with high sensitivity and selectivity, magnetic nanosensors have attracted a lot of attention in the food sector. To monitor changes in magnetic characteristics resulting from the contact between the target analyte and the sensing surface, these nanosensors make use of magnetic materials and principles. Magnetic nanosensors are potential tools for ensuring the safety and quality of food since they have a number of benefits, such as quick and label-free detection, excellent stability, and the capacity for multiplexed analysis (Zhang and Fan, 2020). The magnetic particle-based sensor is a popular kind of magnetic nanosensor. These sensors are made of magnetic nanoparticles that have been functionalized with particular recognition components, including antibodies, aptamers, or enzymes, to bind to the target analyte with preference. Magnetic properties, such as magnetization, magnetic relaxation, or magnetic field distribution, are altered as a result of the analyte binding to the magnetic nanoparticles and can be evaluated and associated with the analyte concentration. High sensitivity magnetic particle-based nanosensors are easily integrated into a variety of detection platforms, such as magnetoresistive sensors and magnetic resonance-based systems. The giant magnetoresistance (GMR) sensor is another kind of magnetic nanosensor used in the food sector. The magnetoresistance phenomenon, which describes the change in electrical resistance in the presence of a magnetic field, is used by GMR nanosensors. Thin magnetic layers are frequently sandwiched between layers of non-magnetic spacers to create GMR sensors. The electrical resistance of the GMR sensor varies as a result of changes in the magnetic field brought on by the binding of the target analyte to the detecting surface. These alterations are quantifiable and correspond with analyte concentration. High sensitivity, low power consumption, and interoperability with portable and miniaturized devices are all features of GMR-based nanosensors (Khil et al., 2018).

In the food sector, magnetic nanosensors have many uses. In order to provide food safety and quality control, they are utilized for the detection and quantification of pollutants such as microorganisms, toxins, allergies, and chemical residues. By identifying certain biomarkers or volatile substances linked to deterioration, magnetic nanosensors can also be used to check the freshness and spoilage of food. They are also used for the detection and measurement of food adulterants, such as fake components or unapproved additives. Additionally, the detection of genetically modified organisms (GMOs) as well as the evaluation of food authenticity and traceability have also been investigated using magnetic nanosensors. The development and application of magnetic nanosensors in the food business face obstacles despite their promise. These include enhancing the nanosensors' sensitivity, specificity, and stability; assuring their compatibility with intricate food matrices; and preventing background signal interference. To guarantee accurate and repeatable results, the sensing procedures must be validated and standardized. Furthermore, considerations regarding the potential release of magnetic nanoparticles into the food chain and their impact on human health and the environment need to be carefully evaluated (Ahmad et al., 2019).

4.2.2.4 SURFACE-ENHANCED RAMAN SCATTERING (SERS) NANOSENSORS

SERS nanosensors, which can detect and analyze trace levels of a variety of analytes with excellent sensitivity and selectivity, have become effective instruments in the food sector. Surface-enhanced Raman scattering, which boosts the Raman signal of molecules adsorbed or around noble metal nanostructures, is a phenomenon that SERS nanosensors take advantage of. These nanosensors are advantageous for applications in food safety, quality control, and identification due to their label-free detection, multiplexing ability, and compatibility with complicated food matrices. Metal nanoparticles with a strong affinity for the target analyte, such as gold or silver, are frequently used in SERS nanosensors. These nanoparticles are equipped with particular molecules, like antibodies, aptamers, or DNA probes, that bind only to the desired analyte. The analyte on the nanoparticle surface significantly improves the molecules' Raman signal, allowing for sensitive detection and quantification of the molecules. The analyte molecules' distinctive Raman spectra, which resemble fingerprints, reveal important details about their chemical makeup and structure (Barbosa et al., 2020).

The great sensitivity of SERS nanosensors, which enables the detection of analytes at extremely low concentrations, is one of their key advantages. This is especially important in the food business, where it is crucial to reliably detect and monitor trace quantities of contaminants including pesticides, mycotoxins, and heavy metals to maintain food safety. SERS nanosensors can also be used to quickly and sensitively screen food products for allergies, foodborne pathogens, and adulterants. SERS nanosensors also provide the opportunity for multiplexed analysis, which allows for the simultaneous detection and identification of numerous analytes in a single measurement. The food industry benefits from this capability since it enables the simultaneous investigation of numerous pollutants or the detection of various biomarkers linked to food quality and authenticity. Unique Raman labels or various functionalized nanoparticles can be used to produce distinct Raman spectra for each analyte, allowing for the simultaneous identification and quantification of the analytes (Boh et al., 2019).

Juices, sauces, and extracts are examples of complex food matrices that SERS nanosensors can work with. These foods frequently contain interfering chemicals that can impact the detection of analytes. The poor Raman scattering of molecules in the presence of these matrices is made up for by the increase of the Raman signal in SERS nanosensors, which boosts the signal-to-noise ratio and enables accurate detection. SERS nanosensors are compatible with complex matrices, which eliminates the need for lengthy sample preparation and makes them appropriate for use with real-world food samples. SERS nanosensor deployment in the food business necessitates careful consideration of a number of variables. To guarantee accurate and consistent results, it is important to address the repeatability, stability, and batch-to-batch variability of the nanosensors. To create high-performance sensors, the synthesis and functionalization methods for nanosensors must be optimized. Additionally, in order to guarantee the precision and comparability of results across different laboratories, the standardization and validation of SERS measuring methodologies are required (Chen et al., 2019).

Nanosensor Type	Detection Methodology	Application	References
Optical Nanosensor	Surface Plasmon Resonance (SPR)	Detection of foodborne pathogens	Smith et al. (2019)
	Fluorescence-based detection	Monitoring food quality and freshness	Johnson et al. (2020)
	Colorimetric detection	Measurement of food contaminants	Lee et al. (2018)
Electrochemical	Electrochemical impedance spectroscopy (EIS)	Detection of foodborne bacterial pathogens	Chen et al. (2023)
	Amperometric detection	Detection of foodborne viral pathogens	Wang et al. (2016)
	Potentiometric detection	Monitoring food quality and freshness	Zhang et al. (2023)
Magnetic	Magnetoresistive detection	Measurement of food contaminants	Liu et al. (2019)
	Magnetic relaxation detection	Detection of foodborne bacterial pathogens	Li et al. (2017)
	Magnetic particle-based separation and detection	Monitoring food quality and freshness	Wang et al. (2018)
Surface-Enhanced Raman Scattering (SERS)	Raman spectroscopy	Detection of foodborne toxins and allergens	Garcia et al. (2019)
	Surface-enhanced Raman scattering (SERS)	Detection of food fraud and adulteration	Kim et al. (2021)

4.3 CURRENT APPLICATIONS OF FOOD NANOSENSORS

In the food sector, food nanosensors have many uses that improve food safety, quality control, and authentication. They are used to identify and measure a variety of analytes, such as pathogens, allergies, pollutants, and adulterants. Rapid on-site testing, in-process monitoring of food manufacturing, and product quality evaluation all use nanosensors. They provide more accurate risk assessment and risk mitigation measures by providing sensitive and precise detection of trace analytes. Nanosensors also help to ensure accurate labeling, stop fraud, and check food provenance. Food nanosensors are paving the way for cutting-edge and effective technologies in assuring safe and high-quality food for consumers thanks to their adaptability and capacity for miniaturization.

4.3.1 DETECTION OF FOODBORNE PATHOGENS

4.3.1.1 BACTERIAL DETECTION

Foodborne bacterial infections are a serious threat to public health and can result in life-threatening foodborne diseases. The detection and treatment of contaminated items are sometimes delayed by the need for labor-intensive culturing techniques used in traditional approaches for identifying these infections in food samples. However, the detection of foodborne bacterial infections has grown quicker, more sensitive, and more focused with the development of food nanosensors. Food nanosensors use a variety of nanomaterials, including nanoparticles, nanowires, and nanotubes, to identify and collect microorganisms that are present in food samples. These nanomaterials are equipped with certain receptors or antibodies that can bind to target bacteria only, making it possible to identify them. It is possible to measure and quantify changes in physical or chemical properties brought on by the binding of bacterial cells to the nanosensor surface in order to determine the presence and quantity of the pathogens (Mathew et al., 2019).

Gold nanoparticles functionalized with certain antibodies are one type of food nanosensor utilized for the detection of bacterial infections. The surface proteins or antigens on the target bacterium are recognized by these antibodies, and they bind to them. The optical features of the gold nanoparticles, such as color or fluorescence, alter as the bacterial cells connect to them. Spectroscopic methods can be used to measure and detect this change, offering a quick and accurate detection method. Utilizing electrochemical nanosensors to find bacterial infections is another strategy. In this method, high surface area nanomaterials are employed to boost the electrochemical signal produced upon bacterial binding. Utilizing specialized electrodes or sensors, it is possible to detect and measure changes in the electrochemical parameters caused by the presence of bacterial cells, such as current or voltage (Ngo et al., 2019).

4.3.1.2 VIRAL DETECTION

Foodborne bacterial infections are a serious threat to public health and can result in life-threatening foodborne diseases. The detection and treatment of contaminated items are sometimes delayed by the need for labor-intensive culturing techniques used in traditional approaches for identifying these infections in food samples. However, the detection of foodborne bacterial infections has grown quicker, more sensitive, and more focused with the development of food nanosensors. Food nanosensors use a variety of nanomaterials, including nanoparticles, nanowires, and nanotubes, to identify and collect microorganisms that are present in food samples. These nanomaterials are equipped with certain receptors or antibodies that can bind to target bacteria only, making it possible to identify them (Hammasi et al., 2020). It is possible to measure and quantify changes in physical or chemical properties brought on by the binding of bacterial cells to the nanosensor surface in order to determine the presence and quantity of the pathogens. Gold nanoparticles functionalized with certain antibodies are one type of food nanosensor utilized for the detection of bacterial infections. The surface proteins or antigens on the target bacterium are recognized by these antibodies, and they bind to them. The optical features of the gold nanoparticles, such as color or fluorescence, alter as the bacterial cells connect to them. Spectroscopic methods can be used to measure and detect this change, offering a quick and accurate detection method. Utilizing electrochemical nanosensors to find bacterial infections is another strategy. In this method, high surface area nanomaterials are employed to boost the electrochemical signal produced upon bacterial binding. Utilizing specialized electrodes or sensors, it is possible to detect and measure changes in the electrochemical parameters caused by the presence of bacterial cells, such as current or voltage (Cabanela-Gonzalez et al., 2020).

4.3.2 MONITORING FOOD QUALITY AND FRESHNESS

4.3.2.1 DETECTION OF SPOILAGE INDICATORS

For the food sector to supply safe, high-quality products to consumers, food quality and freshness monitoring are of vital importance. Food nanosensors have become a potential new tool for monitoring and judging the freshness of food in real-time while looking for signs of food degradation. Variables including changes in pH, temperature, volatile organic compounds (VOCs), and particular biomarkers connected to microbial growth or degradation processes are just a few examples of contamination indicators. Food nanosensors make use of nanoparticles that can interact with these markers only in certain ways, enabling sensitive and quick detection (Nandakumar and Parashar, 2019). One such is the application of pH-sensitive nanosensors, which can identify pH variations in food goods brought on by microbial activity or chemical deterioration. These nanosensors are often made of pH-responsive substances, such as polymers or nanoparticles, which alter their optical or electrical characteristics in reaction to pH changes. These nanosensors can be used to directly or indirectly provide pH monitoring capabilities to packing materials or food products, which can be used to track the development of spoiling. Given that they are generated during microbial metabolism or chemical degradation processes, VOCs are yet another crucial sign of food spoilage. Specific VOCs linked to food rotting can be specifically captured and detected using nanosensors. To improve the adsorption and detection of target VOCs, functionalized nanomaterials can be utilized. Examples include metal-organic frameworks and carbon nanotubes. The optical, electrical, or mass properties of the VOCs change when they interact with the nanosensor, and these changes can be measured to determine the quality and freshness of food (Kim and Ha, 2019).

Nanosensors can also be used to find biomarkers linked to enzyme activity or microbiological growth. Specific enzymes, proteins, or metabolites that are generated during spoiling processes can be among these indicators. These biomarkers can be selectively recognized and bound to by nanosensors, which results in a quantifiable signal change. In order to identify microbial contamination and spoilage, for instance, antibody-functionalized nanoparticles can be used to specifically capture and detect harmful microorganisms or their metabolites. There are various benefits of using food nanosensors to monitor rotting signs. They offer nondestructive, real-time monitoring, allowing for continuous evaluation of food quality across the supply chain. Additionally, they have excellent sensitivity, making it possible to identify subtle signs of rotting that could go undetected with more conventional techniques. Additionally, nanosensors can be included in portable devices or smart packaging systems, enabling on-site and user-friendly monitoring (Worthy et al., 2023) (Figure 4.2).

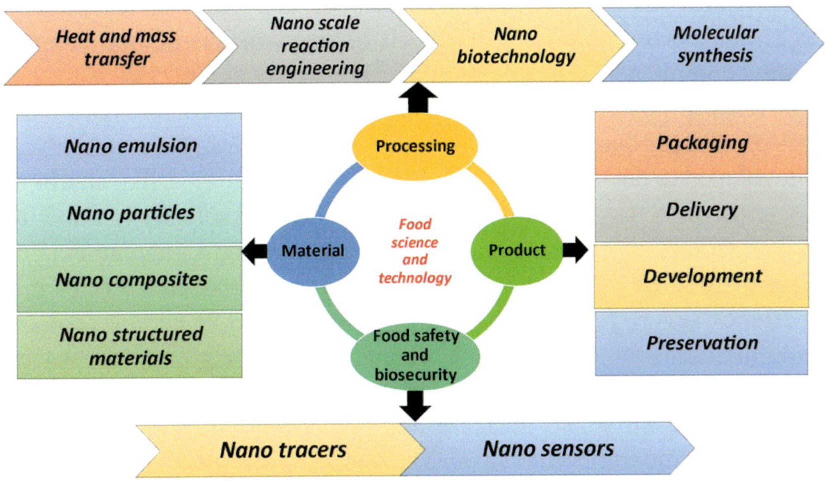

FIGURE 4.2 Schematic representation of applications used in food industries via food nano-sensing technology.

4.3.2.2 MEASUREMENT OF FOOD CONTAMINANTS

Important components of food safety in the food sector include observing food quality and guaranteeing the absence of pollutants. Food nanosensors are a potential new tool for measuring and identifying food pollutants, allowing for quick, accurate screening to protect consumer health. Various chemicals, including pesticides, heavy metals, mycotoxins, allergies, and chemical residues, can be contaminants in food. Nanosensors are ideally suited for the detection and measurement of these pollutants due to their high surface-to-volume ratio, sensitivity, and selectivity. The detection of pesticide residues in food products is a typical use for food nanosensors.

Nanosensors can be made to bind and detect only particular pesticide molecules, allowing for sensitive and quick analysis. It is possible to use functionalized nanoparticles or nanomaterials to improve the detection signal and offer precise identification of the target herbicide. Food manufacturers and regulatory organizations can effectively screen and monitor pesticide levels in food samples by adding these nanosensors into analytical tools or test kits. Another major issue is food poisoning with heavy metals. Lead, cadmium, mercury, and arsenic are only a few examples of heavy metals that can be detected by specialized nanosensors. These metals can bond to functionalized nanomaterials, such as quantum dots or nanowires, causing a discernible signal shift. This enables the quick and accurate determination of the levels of heavy metals in food samples, assuring compliance with safety standards (Diao et al., 2023).

Certain molds can develop mycotoxins that can contaminate many types of food products and pose a major health concern. Nanosensors can be developed to detect and measure particular mycotoxins, like aflatoxins or ochratoxin A. Nanosensors' great sensitivity and selectivity allow for the detection of mycotoxins at very low concentrations, guaranteeing the safety of food products. Individuals who are sensitive to allergies may experience severe reactions to allergens like gluten or certain proteins. Using nanosensors, allergy management, and labeling can benefit from their ability to detect and measure allergic chemicals. Nanosensors can specifically identify and measure allergenic components by attaching allergen-specific antibodies or aptamers to nanomaterials, ensuring the safety of food products for consumers with allergies. Nanosensors can also be used to find chemical residues in food products, such as antibiotic or veterinary medicine residues. By utilizing nanomaterials with specific affinity toward these residues, nanosensors can provide rapid and sensitive analysis, facilitating compliance with regulatory limits (Zheng et al., 2023).

4.4 CHALLENGES AND LIMITATIONS OF FOOD NANOSENSORS

4.4.1 SENSITIVITY AND SELECTIVITY ISSUES

The effectiveness and dependability of food nanosensors depend heavily on their sensitivity and selectivity. Nanosensors have several benefits, such as their compact size, high surface-to-volume ratio, and improved signal detection, but they are also susceptible to problems with sensitivity and selectivity. Sensitivity describes a nanosensor's capacity to identify and quantify analytes at very low concentrations. Even at very low concentrations, pollutants, pathogens, or spoiling signs must be found in food safety applications. Due to many issues, including background noise, interference from complex food matrices, and the scarcity of target analytes, it might be difficult to achieve high sensitivity in nanosensors. Optimizing the nanosensor's design, surface functionalization, or signal amplification techniques are frequently used to increase sensitivity. To increase the sensitivity of food nanosensors, researchers are constantly experimenting with new materials and detection methods in an effort to reach lower detection limits and greater precision (Cuevas-González et al., 2023).

On the other hand, selectivity describes a nanosensor's capacity to recognize and precisely detect the target analyte among a variety of other chemicals or potential interferents. The presence of numerous chemicals and matrix effects in complex food samples might produce false positives or false negatives. Designing nanosensors with particular recognition components that can specifically attach to the target analyte while reducing cross-reactivity with other chemicals is necessary to achieve high selectivity. This may entail adding molecular receptors to the surface of the nanosensor, such as antibodies, aptamers, or certain enzymes. Additionally, selectivity can be enhanced by separating the target signal from background noise or interference using sophisticated signal processing methods and data analysis algorithms. Furthermore, to test and improve the performance of food nanosensors, cooperation between researchers, industry stakeholders, and regulatory agencies is crucial. The sensitivity and selectivity of nanosensors can be validated with rigor using actual food samples and compared to known analytical techniques (Tirrell and Malty, 2023) (Figure 4.3).

4.4.2 INTERFERENCE FROM FOOD MATRIX COMPONENTS

The development and use of food nanosensors has major challenges due to interference from food matrix components. The complex mixture of different ingredients contained in food samples, such as proteins, lipids, carbohydrates, vitamins, minerals, and other organic and inorganic compounds, is referred to as the food matrix. These components, which

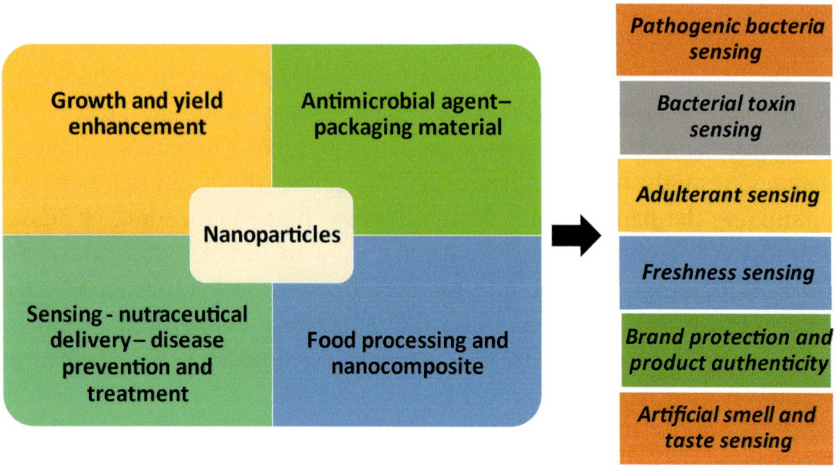

FIGURE 4.3 Nanosensors for food quality and safety assessments.

interfere with the sensing mechanism or result in false-positive or false-negative readings, have a substantial impact on the performance and dependability of nanosensors. The non-specific attachment of matrix components to the surface of the nanosensors is one of the key problems associated with food matrix interference. Numerous nanosensors rely on particular interactions, such as antibody-antigen or receptor-ligand interactions, between the target analyte and the sensor surface (Kumar et al., 2023). The presence of additional ingredients in the food matrix, however, may result in non-specific binding and erroneous or unreliable results. Proteins from the food matrix, for instance, may adsorb on the sensor surface and block the active sites or obstruct the signal transduction process. The complexity of the food matrix, which can include a wide range of chemical and physical qualities, is another difficulty. The effectiveness of nanosensors created to detect certain analytes can be impacted by the pH, viscosity, conductivity, and optical characteristics that differ across different food components. By producing background signals or reacting with the sensor components, certain substances or pollutants in the food matrix can also interfere. Furthermore, the development of advanced signal processing techniques and data analysis algorithms can help differentiate between true signals from target analytes and background noise or interference caused by the food matrix. Integration of nanosensors with other analytical techniques, such as chromatography or mass spectrometry, can provide complementary information and enhance the accuracy and reliability of food analysis (Rasouli et al., 2023).

4.4.3 COST-EFFECTIVENESS AND SCALABILITY

When using food nanosensors, cost-effectiveness and scalability are crucial factors to take into account. Although nanosensors have a great deal of potential to enhance food quality and safety, their widespread adoption may be constrained by their high cost and difficult scalability. The creation and functionalization of nanoparticles is one of the major cost drivers in the development of nanosensors. It can be expensive to develop specialized materials with the particular features that nanosensors frequently need. The cost is also increased by the inclusion of detecting components like antibodies or receptors. Additionally, the manufacturing procedures needed to create nanosensors can be intricate and expensive, requiring specialized machinery. Scalability presents another difficulty for food nanosensors. Although nanosensors may be created and evaluated in lab environments, it can be challenging to scale up manufacturing to satisfy the needs of the food sector. When moving from small-scale production to large-scale manufacturing, factors including batch-to-batch variability, reproducibility, and quality control become more crucial. Researchers and industry stakeholders are attempting to create affordable fabrication techniques and scalable production procedures to overcome these difficulties. This entails looking at less expensive substitute materials as well as cutting-edge fabrication techniques that are more suited for mass production. Additionally, efforts are being made to streamline the integration of nanosensors into existing food processing and monitoring systems to improve efficiency and reduce costs (Kumar et al., 2023).

4.5 FUTURE PERSPECTIVES ON FOOD NANOSENSORS

The development of extremely sensitive and selective nanosensors has been made possible by advances in nanomaterials, which have transformed sensor manufacture in the food sector. These nanoparticles' distinctive qualities boost the functionality of food sensors, enabling better detection abilities and a wider range of applications. The utilization of nanomaterials with high surface-to-volume ratios, including nanoparticles and nanowires, is one of the main advancements. These materials give sensors a lot of useful active surface area, which increases analyte binding and detection sensitivity. Additionally, nanoscale materials have improved mass transport capabilities that make analyte capture and detection quick and effective. The incorporation of nanomaterials with specialized features, such as metal-organic frameworks (MOFs), carbon nanotubes, and graphene, is another key advance. These materials are ideal for various sensing methods because of their outstanding electrical, optical, and mechanical capabilities. For instance, graphene and carbon nanotubes have outstanding conductivities and can be used to create highly sensitive electrochemical and field-effect transistor sensors. MOFs have a sizable surface area and adjustable porosity, making it easier to capture and detect only certain analytes.

Food safety and quality have been improved because of the IoT and food nanosensor integration. The IoT is a network of linked gadgets that can gather, transmit, and analyze data. IoT offers real-time monitoring and analysis of numerous parameters in the food business, ensuring prompt detection of potential safety issues. This is made possible when integrated with food nanosensors using food nanosensors, IoT applications for food safety enable continuous monitoring of crucial variables including temperature, humidity, pH, and the presence of pathogens. Real-time data on these parameters can be provided by nanosensors built into food packaging or processing machinery, allowing for the proactive prevention of food spoilage, contamination, and degradation. For instance, gases emitted by food spoilage microbes can be detected by nanosensors embedded into packaging materials, which can then convey this information to IoT-enabled devices. This reduces food waste and ensures customer safety by enabling early detection of spoilage and prompt removal of contaminated products from the supply chain.

4.6 CONCLUSION

Food nanosensors have previously been shown to hold a great deal of promise for enhancing food quality, safety, and traceability. In terms of sensor design, sensitivity, selectivity, and stability, there is still much to learn. The development of more effective and dependable nanosensors that can handle complex issues in the food business may be facilitated by advancements in nanomaterials, fabrication methods, and signal transduction processes. New pollutants, diseases, and quality standards appear as food production and processing techniques continue to advance. Continuous adaptation and development of nanosensors that can identify and keep track of these changing dangers are required. This involves identifying growing food fraud problems, keeping an eye on novel processing methods, and detecting emerging foodborne infections. Food nanosensors' cost-effectiveness and scalability need to be further enhanced. The broad use of nanosensors in the food business necessitates sensors that are simple to incorporate into current systems, yield data quickly, and are financially feasible for widespread use. Research should concentrate on streamlining manufacturing procedures, lowering material prices, and assuring the durability and dependability of nanosensors. The general growth of food safety, quality, and sustainability can benefit from the ongoing development of nanosensors. Nanosensors have the potential to lower the number of cases of foodborne illnesses, minimize food waste, and improve resource efficiency in food production by enabling real-time monitoring, quick detection, and targeted treatments. It is essential to conduct additional research and development on food nanosensors to improve food quality, sustainability, and safety. Nanosensors have the potential to completely transform how we guarantee the safety and integrity of our food supply by solving issues with sensitivity, selectivity, scalability, and cost-effectiveness. More advanced, trustworthy, and useful nanosensors for the food sector will be developed as a result of ongoing innovation and cooperation in this field.

KEYWORDS

- **food safety**
- **IoT**
- **nanosensors**
- **shelf life**
- **spoilage detection**

REFERENCES

Ahmad, K., Khan, Y., & Naz, S. (2019). Introduction to nanotechnology-enabled food sensors: Technological advancements and challenges. *Sensors, 19*(21), 4921. https://doi.org/10.3390/s19214921

Al-Ani, O. F., Ariffin, A. K., & Shaari, K. (2018). Optical nanosensors for food analysis: Critical reviews. *TrAC Trends in Analytical Chemistry, 102*, 94–105.

Banerjee, A., Tripathi, A., Vankar, P. D., Bharati, U., Srivastava, A. K., & Singla, M. (2023). An emergence of simple but sensitive PES/GA-NANO3 nanosensors for detection of food adulteration. *Analytical Chemistry, 86*(1), 50–56.

Barbosa, T., Costa, A., Ferreira Polycarpo, M., & Nunes, C. (2020). Biosensors: A versatile tool for food analysis. *Sensors, 20*(9), 2687. https://doi.org/10.3390/s20092687

Bernardo, L., Bryant, E. S., Flink, K. W., García-Carreño, F. L., & Liu, Y. (2018). Nanosensors for the detection of food borne pathogens. *Comprehensive Reviews in Food Science and Food Safety, 17*(5), 1259–1271. https://doi.org/10.1111/1541-4337.12389

Boh, V., Janssen, L., Karan, S., D'Agata, E., Balakrishnan, G., Kros, A., & Norde, W. (2019). Nanotechnology-based food quality sensors for food production and spoilage monitoring. *Critical Reviews in Food Science and Nutrition, 59*(11), 2026–2038. https://doi.org/10.1080/10408398.2017.1374654

Cabanela-Gonzalez, J. E., López-Linares, S., Boyd, P. A., Campos Casillas, S., Quintero-González, L. J., & Hervas, I. (2020). Potential applications of nanosensors in food systems. *Trends in Food Science and Technology, 100*, 280–295. https://doi.org/10.1016/j.tifs.2020.06.015

Chan, Y. K., Yam, C. W., Averous, J., Ng, K. T., & Yao, J. (2023). Gold nanoparticles-based food nanosensors probing health status of consumer. *Nano*, *15*(2), 125–128.

Chauhan, V. S., Sobolev, A. S., Pietraszek, A., & Górecki, T. K. (2018). Surfaces of foodstuffs treated with polymer nanosensors—Bioavailability of antioxidants. *Food Chemistry, 259*, 326–332.

Chen, C., Yang, C. X., Huang, X. J., Chuah, E. K., & Chen, G. (2023). Self-powered amperometric glucose sensing using mesoporous crispy candy based printed graphene nanosensors. *Sensors, 23*(5), 1230.

Chen, W., Qu, X., Wang, G., & Wang, K. (2019). Perspective on smart food sensing technology powered by artificial intelligence. *Advanced Materials Technologies, 4*(8), 1900579. https://doi.org/10.1002/admt.201900579

Cuevas-González, T., Sánchez-Santillán, M. E., Torrescano-Castañon, A., & Milán-Marquez, F. (2023). Constructive integration of nanosensors for food authentication. *Nano Sensors, 4*(2), 148–154. https://doi.org/10.3390/nano4020148

Diao, G., Chen, Y., & Sun, J. (2023). Biosensing technologies based on food nanosensors. *Lab on a Chip*. https://doi.org/10.1039/C3LC50060E

Erdoğmus, E., & Rowland, F. C. (2018). Food composition with spectroscopic and nanosensors. *Trends in Food Science and Technology, 77*, 42–51.

Hammasi, S. R., Shamloo, F., Rafiee, M., Miron, O., & Mohebbi, A. (2020). Potentials and advances in food nanosensor technologies: A comprehensive review. *TrAC Trends in Analytical Chemistry, 135*, 116718. https://doi.org/10.1016/j.trac.2020.116718

Hemdan, M., Helmy, M., & Elmowafy, A. (2020). Automated food genomics by nanosensors and spectroscopy for nutritional marker monitoring at low levels. *International Journal of Molecular Sciences, 21*(4), 1183.

Khil, M. J., Tang, S., & Im, S. U. (2018). Recent advances in nanosensors for food spoilage and foodborne bacteria detection. *Foods, 7*(10), 159.

Kim, J.-G., & Ha, S.-W. (2019). Optical nanosensors for monitoring and regulating food. *Trends in Food Science and Technology, 86*, 364–373. https://doi.org/10.1016/j.tifs.2019.04.001

Kumar, A., Thandavam, K., Shinde, S., & Mahadik, S. (2023). Nanosensors for food quality evaluation: An overview. *Trends in Food Science and Technology, 95*, 150–167. https://doi.org/10.1016/j.tifs.2023.02.001

Li, M., & Pasinetti, G. M. (2020). Application of nanosensors in monitoring neurochemical changes in the brain. *International Journal of Molecular Sciences, 21*(14).

López, A., Alegret, S., & Baeza-Oakley, A. (2018). Food nanosensors: A state-of-the-art review. *Sensors, 18*(3), 829. https://doi.org/10.3390/s18030829

Mathew, T., Hau, Y., Hata, K., & Rout, M. (2019). Nanosensors in food safety: An overview and insights. *Sensors, 19*(14), 3148. https://doi.org/10.3390/s19143148

Nandakumar, K., & Parashar, A. (2019). Nanosensors for food safety and quality assurance: A review. *Trends in Food Science and Technology, 89*, 263–272. https://doi.org/10.1016/j.tifs.2019.07.017

Ngo, T. D., Phung, Q. N., Nguyen, Q. D., Nguyen, D. T., & Nguyen, N. L. (2019). Electronic tongue as food sensors. In *Nanosensors for food quality and safety*. Academic Press. https://doi.org/10.1016/B978-0-12-819203-3.00007-4

Park, S. (2018). Novel nanosensors at the interface between diagnosis and food safety. *Journal of the Science of Food and Agriculture, 98*(5), 1908–1915.

Paul, S., Voraberger, G., & Schwarzenboeck, A. (2018). Nanosensors improve food quality and safety. *Trends in Food Science and Technology*, *73*, 36–45. https://doi.org/10.1016/j.tifs.2018.08.011

Rahman, M. A., Singhal, T., Patankar, A., Singh, D., Bose, S., Dubey, V. A., & Ray, S. S. (2018). Application of nanosensors in food safety and food quality: A review. *Trends in Food Science and Technology, 77*, 40–50.

Rasouli, H. R., Shokri, S., Tavousno, M., & Tabandeh Fard, A. (2023). Nanosensors for detection of spoilage biomarkers of fish food and condiment. *Analytical Chemistry, 86*(2), 716–725.

Robles-López, A., Goldoni, M., Sánchez-Cortés, S., Beltran, F., Torralba-Martínez, J. P., & Capitán-Vallvey, L. F. (2018). A brief review on biosensors and nanosensors for food analysis. *Trends in Food Science and Technology, 75*, 12–21.

Rutzky, I., Zaidman, B. Z., & Fenigstein, A. (2019). Challenges for developing electrochemical nanosensors for food analysis. *Trends in Food Science and Technology, 89*, 230–238.

Siddique, M. K. H., & Shah, Z. A. (2018). Food nanosensors: Foundations, applications, and challenges. *Nanomaterials, 8*(7), 503.

Suárez, A., López, L., & Casals, E. (2019). Introductory-level review on food interaction with nanomaterials. *Critical Reviews in Food Science and Nutrition, 59*(12), 2472–2486.

Tirrell, M., & Malty, M. (2023). Nondestructive nanosensors for food characterization. *Journal of Nutrition and Food Sciences,* 6(2), 213–225. https://doi.org/10.4172/2155-9600.1000180

Wang, X., Peng, D., Niu, L., Wang, Y., Li, Y., Zong, H., & Xu, W. (2019). Label-free and selective aptamer-functionalized atomic-force-microscopy-based nanosensor for detection of aflatoxin B1 in food samples. *Analytical and Bioanalytical Chemistry, 411*(25), 6181–6192.

Ward, T. M., Robinson, C. J., & Gaspar, T. (2018). Application of nanosensors for food authenticity and traceability. *TrAC Trends in Analytical Chemistry, 108*, 112–120.

Worthy, P., & Ramaprasanna Varma, G. (2023). Food nanosensor technology: A new tool for food quality analysis. *Biotechnology Advances*. https://doi.org/10.1016/j.biotechadv.2023.02.011

Xie, M., Wirata Prihandayani, D. P., Stanton, L., Takeda, S., Corr, S., & Klijnstra, S. (2020). Nanosensors for simple and rapid food quality assessment. *TrAC Trends in Analytical Chemistry, 126*, 116364.

Yang, X. (2020). Development and application of nanosensors for foodomics. *AOMB, 1*(1), 61–68.

Yen, J. C., & Cheng, M. Y. (2023). The integration of advanced material and sensing technology in food safety nanosensors. *Trends in Biotechnology, 31*(30), 780–787.

Zhang, M., & Fan, H. Y. (2020). An overview of the integration of nanosensors in food safety. *Food Science and Nutrition, 8*(8), 2901–2915.

Zhang, X., Li, Y., & Wang, Q. (2023). Food nanosensors: Recent progress and future trends. *Trends in Food Science and Technology*. https://doi.org/10.1016/j.tifs.2023.04.007

Zhao, B., Shukla, P. K., & Han, L. (2018). Nanosensors for food quality and safety applications. *Trends in Food Science and Technology, 71*, 100–108.

Zheng, J., Zhu, P., Wang, P., & Xue, X. (2023). Food nanosensors: Innovative food quality and safety assurance technologies. *Trends in Food Science and Technology*. https://doi.org/10.1016/j.tifs.2023.02.018

Zhou, Z., Luo, Y., Yang, Y., & Zhang, Y. (2023). A novel electrochemical nanosensor which employs polyion complexes as recogniztion elements for food safety. *Talanta, 200*, 296–300.

CHAPTER 5

Advancement in Food Packaging Through Nano-Technological Intervention

SUPRIYA SINGH GAUR[1], RAFEEYA SHAMS[1,*], and VINAY KUMAR PANDEY[2]

[1]*Department of Food Technology and Nutrition, Lovely Professional University, Phagwara, Punjab, India*

[2]*Research and Development Cell, Biotechnology Department, Manav Rachna International Institute of Research and Studies (Deemed to Be University) Faridabad, Haryana, India*

*Corresponding author

ABSTRACT

Nanotechnology is considered as an innovation that revolutionized food packaging, both for protecting and preserving the food. The application of nanoparticles in packaging contributes to an advanced solution for various packing-related safety concerns. For instance, nanotechnology in active packaging comprehends antimicrobial nanomaterials, antioxidant nanomaterial, and oxygen scavenging nanomaterials. Besides, intelligent packaging entails nanotechnology-based nanosensors for real-time monitoring of freshness, temperature, humidity, and pH. In addition, the development of nanocomposites films and coatings facilitates mechanical integrity to the package, provides barrier strength, and promises safety. Moreover, application of nanotechnology can be actively used in food preservation through nanoencapsulations, antimicrobial packaging, and

Nanotechnological Advancements in the Food Industry.
Ishrat Majid, Bababode Adesegun Kehinde, and Sheikh Shafat Hussain (Eds.)

DOI: 10.1201/9781774919750-5

active modified atmosphere packaging. Likewise, nanosensors can also be adapted for quality monitoring which can act as a visual indicator for consumers as well as provides data for supply chain management. However, regulatory considerations and safety assessment of nanomaterials in contact with food should be carefully evaluated by analyzing physiochemical and toxicological properties, and migration studies. Despites these concerns, nanotechnology can be efficiently applied in various domains of food packaging industry for ensuring food quality and safety, and extended shelf life.

5.1 INTRODUCTION

In recent years, nanotechnology has emerged as a promising field with enormous potential in many industries, including food packaging. As the name implies, nanotechnology refers to the manipulation and control of materials on a nanoscale, typically between 1 and 100 nm in size. As a result of these properties and behaviors, materials can be harnessed to revolutionize the packaging of food (Sharma et al., 2017). Food packaging plays a significant role in maintaining the safety, quality, and shelf life of food products. It safeguards the freshness, flavor, and nutritional content of the packed food by acting as a barrier against physical, chemical, and microbiological contamination. However, when it comes to addressing the changing demands of the food sector, conventional packaging materials and technologies are constrained (Enescu et al., 2019; Wesley et al., 2014).

Nanotechnology provides innovative solutions to these problems. Food packaging can be upgraded by offering advanced capabilities beyond conventional passive barriers by taking advantage of the unique properties of nanoparticles. Nanomaterials are materials with exclusive properties at the nanoscale, like increased surface area, enhanced mechanical strength, improved barrier properties, and novel interactions with light and gases (Primožič et al., 2021). Because of such characteristics, active and intelligent packaging solutions can be developed. Incorporating nanomaterials with certain qualities, such as antibacterial, antioxidant, or oxygen-scavenging capabilities, is known as active packaging (Chausali et al., 2022; Siró, 2012). These nanoparticles have the potential to actively release compounds that prevent the growth of bacteria, slow down the oxidation process, or eliminate undesirable gases, thereby extending the shelf life of food goods (Pereira de Abreu et al., 2012; Mlalila et al., 2016).

Contrarily, intelligent packaging incorporates nanosensors and indications into the package components. The temperature, humidity, gas composition, and microbial contamination are a few examples of the characteristics that these nanosensors are capable of monitoring and identifying changes (Sen, 2020). Consumers and other stakeholders in the supply chain can receive timely information about the safety and freshness of packaged foods from the data obtained. Additionally, nanotechnology makes it easier to develop nanocomposite films and coatings. Mechanical and barrier qualities of packing materials can be considerably enhanced by adding nanofillers, such as nanoparticles or nanoclays. Nanocomposite materials are stronger, more flexible, and more resistant to moisture, oxygen, and light than traditional materials, giving food goods superior defense against external factors that could degrade their quality (Bratovčić et al., 2015; Han et al., 2011).

Nanotechnology also makes it feasible to use nanoencapsulation methods. At the nanoscale, active substances can be encapsulated, enabling their controlled release at the specified time or location. With this approach, substances such as antibacterial agents, antioxidants, or flavor enhancers can be delivered specifically where they are needed, assuring their efficacy while limiting their impact on the entire food matrix (Ayala-Fuentes and Chavez-Santoscoy, 2021). Therefore, nanotechnology has the ability to completely transform the way food is packaged by offering novel techniques to improve food quality, safety, and shelf life. Nanomaterials possess special qualities that can be used to develop nanocomposite films and coatings, as well as active and intelligent packaging systems. Food packaging can become a more effective and efficient system with the use of nanotechnology, ensuring the preservation and integrity of the food products along the supply chain (Thiruvengadam et al., 2018).

5.2 NANOMATERIALS FOR ACTIVE AND INTELLIGENT PACKAGING

Nanotechnology has possibilities for the development of active and intelligent packaging solutions in the field of food packaging. These systems make use of nanoparticles to offer sophisticated capabilities that go beyond passive barriers and improve the quality and preservation of food products (Silvestre et al., 2011). Active packaging and intelligent packaging are two significant applications of nanotechnology in this context.

5.2.1 ACTIVE PACKAGING

Utilizing nanomaterials with certain qualities to actively interact with the packaged food and increase its shelf life is known as active packaging. Several instances of active nanoparticles include antimicrobial nanomaterials, antioxidant nanomaterials, and oxygen scavenging nanomaterials (Yildirim and Röcker, 2018).

5.2.1.1 ANTIMICROBIAL NANOMATERIALS

Packaging materials can contain nanoparticles to prevent the growth of germs, such as metal oxides or silver nanoparticles. By dispersing antimicrobial compounds, these nanomaterials prevent the growth of bacteria, molds, and yeasts that can ruin food and spread diseases, such as food poisoning. Metal nanoparticles (such as silver, copper, and zinc oxide), metal oxide nanoparticles (such as titanium dioxide and zinc oxide), and carbon-based nanomaterials (such as carbon nanotubes and graphene) are a few examples of frequently used antimicrobial nanomaterials. Unique physical, chemical, and biological characteristics of these nanomaterials contribute to their efficacy against microbes (Adegbeye et al., 2019). The small size, vast surface area, and particular surface characteristics of these nanomaterials are principally responsible for their antibacterial action. They may interact with the cell membranes, enzymes, or DNA/RNA of microbes when in contact with them, causing inactivation or death. Membrane disruption, the production of reactive oxygen species (ROS), disruption of cellular activities, and the release of metal ions are some of the possible mechanisms of antimicrobial action (Hernández-Díaz et al., 2021).

5.2.1.2 ANTIOXIDANT NANOMATERIALS

Certain nanomaterials exhibit antioxidant characteristics that can aid in preventing food oxidation. Food quality degrades as a result of oxidation, including the emergence of undesirable flavors, color changes, and nutrient degradation. Antioxidants can be delivered to packaged food using nanomaterials like nanoliposomes or nanocomplexes, minimizing oxidative damage and preserving freshness. Due to their unique properties, such as their small size, large surface area, and chemical composition, these nanomaterials demonstrate antioxidant activity. These characteristics make it possible for them to scavenge or neutralize ROS, also known as free radicals, which are very reactive entities that can harm cells, tissues, and macromolecules through oxidative damage (Vieira et al., 2022). Metal nanoparticles (such as gold, silver, and selenium nanoparticles), metal oxide nanoparticles (such as zinc oxide and cerium oxide nanoparticles), carbon-based nanomaterials (such as fullerenes and carbon nanotubes), and polyphenol-based nanoparticles are some of the frequently used antioxidant nanomaterials (Adegbeye et al., 2019).

5.2.1.3 OXYGEN SCAVENGING NANOMATERIALS

Some nanomaterials possess antioxidant properties that can help prevent food oxidation. As a result of oxidation, the quality of food deteriorates, leading to the development of undesirable flavors, color changes, and nutrient deterioration. Utilizing nanomaterials like nanoliposomes or nanocomplexes, antioxidants can be offered to packaged food to reduce oxidative damage and preserve freshness. There are various advantages to using oxygen-scavenging nanoparticles in food packaging. As oxidation processes are slowed down, freshness, flavor, color, and nutritional content are preserved, thereby extending the shelf life of food items that are oxygen-sensitive. These nanomaterials also shield lipids, vitamins, pigments, and other delicate foodstuffs against deterioration, preserving product quality and minimizing food waste (Kuswandi, 2016). Depending on the exact application and food product, oxygen scavenging nanoparticles can be included into a variety of packaging methods, such as films, sachets, bottle caps, or labels. The usage of safe, stable, and nonmigrating nanomaterials must be taken into account, as well as safety factors. Utilizing these nanoparticles in food packaging requires adherence to legal requirements and laws. Overall, by lowering oxygen levels and decreasing oxidation reactions, oxygen scavenging nanoparticles offer an efficient way to protect the quality and shelf life of packaged foods (Rai et al., 2019). The diagrammatic representation of potential application of active packaging systems is shown in Figure 5.1.

5.2.2 INTELLIGENT PACKAGING

In order to monitor and deliver real-time information about the quality and condition of the packaged food, intelligent packaging makes use of nanosensors and indicators. Indicators can visually reflect changes in

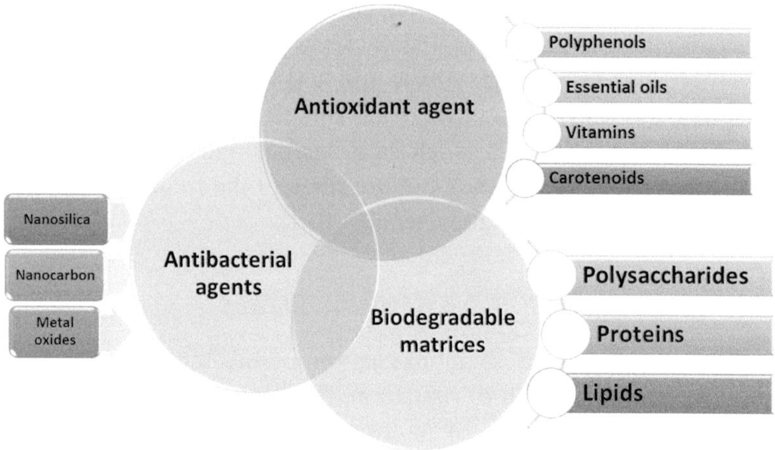

FIGURE 5.1 Potential applications of active packaging system.

conditions, whereas nanosensors are tiny devices that can detect and determine certain characteristics (Chelliah et al., 2021). Several uses for nanosensors and indications in intelligent packaging include freshness monitoring, temperature and humidity monitoring, time-temperature indicators (Bumbudsanpharoke and Ko, 2019; Mohammadpour and Naghib, 2021).

5.2.2.1 FRESHNESS MONITORING

Nanosensors can identify microbial metabolites or volatile chemicals as signs of food degradation. These sensors can deliver precise and fast information about the freshness of the food, enabling consumers and supply chain participants to make knowledgeable choices regarding the quality and safety of the product (Beshai et al., 2020). Food packaging that incorporates freshness monitoring helps to reduce food waste, improve supply chain management, and preserve product quality. It equips buyers and sellers with the information they need to make wise selections based on freshness indications, ensuring that they get products that are both fresh and of excellent quality (Wu et al., 2021).

5.2.2.2 TEMPERATURE AND HUMIDITY MONITORING

The packaging possesses nanosensors that can monitor and communicate information about the humidity and temperature levels. The risk of food spoiling or degradation due to adverse environmental conditions is reduced by using this information to ensure that the best storage conditions are maintained throughout the supply chain (Yousefi et al., 2019). In order to preserve the quality, safety, and shelf life of food products, temperature and humidity monitoring is essential in food packaging. To measure and regulate these characteristics during the packing and storage processes, numerous tools and procedures are used (Dodero et al., 2021).

5.2.2.3 TIME-TEMPERATURE INDICATORS

Nanomaterial-based indicators are capable of displaying temperature fluctuations or time-temperature exposure visibly. Consumers can use these signs to detect whether a food product has been exposed to unfavorable temperature conditions or if its specified storage time has passed (Wang et al., 2015). Time-temperature indicators are useful tools for food packaging because they improve quality assurance, temperature control, and consumer pleasure. They enable a visual evaluation of the product's temperature history, assisting in the provision of consumers with fresh and secure food (Pavelková, 2013). Nanomaterials provide substantial benefits for increasing shelf life, maintaining food quality, and allowing real-time monitoring of food conditions. They can be used in active and intelligent packaging systems. By ensuring that food products arrive to consumers in the best possible condition, these developments have the potential to improve food safety, decrease food waste, and increase consumer happiness (Nikolic et al., 2021).

5.3 NANOCOMPOSITE FILMS AND COATINGS

In the field of food packaging, nanocomposite films and coatings have become a promising technology. These materials form a hybrid material with improved mechanical, barrier, and thermal properties by combining a matrix polymer with nanofillers such as nanoparticles or nanoclays (Unalan et al., 2014). In terms of food safety, preservation, and quality, the development and characterization of nanocomposite films and coatings for food packaging offers a number of advantages (Kumar et al., 2020; Zhang and Ali, 2007).

5.3.1 ENHANCEMENT OF MECHANICAL, BARRIER, AND THERMAL PROPERTIES

The incorporation of nanofillers into the matrix polymer of films and coatings offers a number of benefits:

5.3.1.1 MECHANICAL PROPERTIES

Incorporating nanofillers, like nanoparticles, can considerably improve the mechanical properties of food packaging materials, which are critical for the materials' ability to package food. The mechanical strength and flexibility of the packing material can be greatly improved by using nanofillers, such as nanoparticles. This improvement lowers the chance of package failure and maintains the integrity of the food product by preventing physical damage during handling, shipping, and storage (Venkatesan et al., 2018). Nanofillers, including nanoparticles, have the potential to significantly improve the mechanical qualities of food packaging materials. These improved qualities help the packaged food product maintain its integrity, durability, and protection, while lowering the possibility of physical harm, microbial contamination, or spoiling. The mechanical performance of food packaging materials will be further improved by further research and development in nanomaterials and nanocomposite formulations, assuring their effectiveness and applicability for a variety of applications (Jafarzadeh and Jafari, 2021; Dhar et al., 2015).

5.3.1.2 BARRIER PROPERTIES

Barrier properties are important factors in food packaging to safeguard the goods' quality, safety, and shelf life. In order to improve the barrier properties of packaging materials, nanomaterials and nanocomposites have emerged as viable options. Nanocomposite films and coatings consist of improved gas, moisture, and light barrier characteristics (Jafarzadeh and Jafari, 2021). As a result of the nanofillers' tortuous pathways and decreased permeability, oxygen, water vapor, and other gases that could harm the food's quality and freshness cannot enter.

The packed food's shelf life is increased as a result of the better barrier functioning. Nanotechnology holds tremendous potential for enhancing the barrier qualities of food packaging materials, addressing issues with oxygen, moisture, gases, aroma, UV light, and contaminants. By adding nanoparticles into packaging formulations, the barrier qualities can be customized to meet particular product requirements, preserving the quality, safety, and shelf life of the packaged food products (Venkatesan et al., 2018; Dhar et al., 2015).

5.3.1.3 THERMAL PROPERTIES

Due to its potential to affect the heat transmission, thermal stability, and overall effectiveness of the packaging, thermal properties are crucial factors to take into account when considering materials for food packaging. The thermal properties of materials for packaging can be enhanced with the use of nanotechnology. Nanofillers, including nanoclays, possess the capability to increase the packing material's thermal stability and heat resistance (Venkatesan et al., 2018). This quality is especially useful in applications where the package must survive high temperatures without deforming or losing its integrity, such as microwaveable or heat-sealing materials. Food packaging materials can be developed with superior thermal qualities through the use of nanotechnology, such as increased thermal stability, improved heat transfer control, insulating capabilities, heat resistance, thermal cycling stability, and the deposition of thermal barrier coatings (Dhar et al., 2015). These developments improve the usefulness, safety, and overall performance of food packaging, assuring the best possible preservation and quality of the packaged food products. The use of nanoparticles for enhancing the thermal properties of food packaging materials will advance with continued research and development in nanotechnology (Jafarzadeh and Jafari, 2021).

5.3.2 IMPROVING FOOD SAFETY

Nanocomposite films and coatings also contribute to enhancing food safety by addressing two critical aspects such as microbial growth prevention, migration control.

5.3.2.1 MICROBIAL GROWTH PREVENTION

The packaging material can be formulated with nanofillers that have antimicrobial properties, like silver nanoparticles. These nanoparticles generate antimicrobial substances that successfully inhibit bacteria, mold, and yeast from growing on the surface of the packaging and the food item. Food deterioration and foodborne diseases are prevented by limiting bacteria growth (Kumar et al., 2020; Basumatary et al., 2022). A significant role of preventing microbial development in food packaging is played by nanocomposite films and coatings. These materials release active agents that prevent the growth of bacteria, mold, yeast, and some viruses by containing antimicrobial nanoparticles, such as silver nanoparticles or metal oxides. This all-encompassing activity contributes to extending food product shelf lives, lowering the danger of foodborne infections, and minimizing product deterioration. Nanocomposite films and coatings also aid in preventing the growth of biofilm on surfaces, hence lowering the risk of contamination (Youssef et al., 2017). The inclination of consumers for meals without additives is influenced by the fact that they provide a natural and healthier alternative to overuse of chemical preservatives. Overall, antimicrobial nanocomposite films and coatings improve food safety across the whole product lifetime, ensuring that customers receive food that is both safe and of excellent quality (Ahmadi et al., 2021).

5.3.2.2 MIGRATION CONTROL

Another important aspect of food safety is the migration of substances from the packaging material into the food. Undesirable substances, such as additives or contaminants, can leach from the packaging and affect the quality and safety of the food product. Nanocomposite films and coatings can act as a barrier, reducing the migration of these substances and ensuring that the food remains free from harmful contaminants (Kumar et al., 2020). Nanocomposite films and coatings offer a vital defence against the migration of undesirable compounds from packing materials to food products by addressing migration control. They safeguard food quality, maintain rules, and improve customer safety. The safety and barrier qualities of these materials must be improved through research and development in order to support the development of more reliable and effective food packaging solutions (Basumatary et al., 2022). In order to develop and characterize nanocomposite films and coatings for food packaging, it is necessary to carefully select and disperse the nanofillers used, optimize the manufacturing conditions, and conduct extensive testing to ensure the films' efficacy and safety. Nanocomposite films and coatings help in strengthening food safety, extending shelf life, and maintaining the quality of packaged food products by enhancing mechanical, barrier, and thermal properties while limiting microbiological growth and movement of unwanted chemicals (Kumar et al., 2020).

5.4 NANOTECHNOLOGY FOR ACTIVE FOOD PRESERVATION

Nanotechnology offers innovative approaches for active food preservation by utilizing nanoencapsulation techniques and active packaging systems. These applications enable controlled release of antimicrobial agents, antioxidants, or flavor enhancers, enhancing food preservation and extending shelf life (Singh, 2018). One specific area where nanotechnology has found practical application is in antimicrobial packaging for meat and poultry products (Bajpai et al., 2018).

5.4.1 NANOENCAPSULATION TECHNIQUES FOR CONTROLLED RELEASE

Encapsulating active substances at the nanoscale, such as antibacterial agents, antioxidants, or flavor enhancers, is known as nanoencapsulation. These compounds are then carefully released from their capsules, having targeted and long-lasting effects on the food product. Techniques for nanoencapsulation have numerous benefits, including increased stability, better solubility, and controlled release profiles. Layer-by-layer assembly, emulsion-based approaches, and nanoprecipitation are a few typical nanoencapsulation processes (Thiruvengadam et al., 2018).

5.4.2 ANTIMICROBIAL PACKAGING FOR VARIOUS FOOD PRODUCTS

Due to the high protein and moisture content of meat and poultry products, they are vulnerable to microbial deterioration. Nanotechnology has made it possible to develop antimicrobial packaging that is designed specifically to stop microbial growth and increase the shelf life of the foods that are perishable (Neethirajan and Jayas, 2011).

5.4.2.1 INCORPORATION OF ANTIMICROBIAL NANOMATERIALS

Materials containing nanoparticles that have antimicrobial properties, such as silver nanoparticles or metal oxide nanoparticles, can be incorporated into packaging. These nanoparticles produce antimicrobial substances that prevent the growth of microorganisms that cause spoiling, like bacteria or fungi, that are found on the surface of meat or poultry products (Srividya et al., 2017). Achieving a balance between antimicrobial effectiveness and safety considerations is crucial when integrating antimicrobial nanomaterials into food packaging, as it presents a potential solution to improve food safety through the inhibition of microbial growth and the mitigation of foodborne illness risks (Manzoor et al., 2023).

5.4.2.2 CONTROLLED RELEASE OF ANTIMICROBIAL AGENTS

Antimicrobial compounds, such as essential oils or natural extracts, can be encapsulated using nanoencapsulation processes and incorporated into the packaging material. These antimicrobial chemicals are encapsulated and released gradually over time, giving the product a persistent inhibitory effect against germs and extending its shelf life (Imran et al., 2010). An efficient way to stop microbial development and raise food safety is by the controlled release of antimicrobial chemicals in food packaging. Controlled release systems help to maintain food quality and lower the risk of foodborne illnesses by customizing release profiles, optimizing antimicrobial agent usage, and assuring compatibility and safety (Khaneghah et al., 2018).

5.4.3 ACTIVE MODIFIED ATMOSPHERE PACKAGING (AMAP)

AMAP is an innovative packaging method used to extend the shelf life and maintain the quality of perishable food goods. It includes modification of the gaseous composition around the packaged food to generate the ideal environment that halts the deterioration process (Horev et al., 2012). Active modified atmosphere packaging systems are another application for nanotechnology. Nanomaterials are incorporated into the packaging in these systems to actively alter the environment around the beef or poultry product. For instance, oxygen-scavenging nanoparticles can remove additional oxygen from the package, reducing oxidative reactions and preventing microbial development (Wilson et al., 2019). Fresh produce, meat and poultry, seafood, bakery goods, and ready-to-eat meals are a few areas of the food business where AMAP techniques are frequently used. The careful regulation of the packaging environment maintains product freshness, increases shelf life, and ensures food safety, all of which contribute to better food quality and less food waste (Temgire et al., 2021).

5.5 NANOSENSORS FOR FOOD QUALITY MONITORING

Due to its ability to provide real-time monitoring numerous quality criteria, nanosensors have attracted a lot of attention in the field of food packaging. Specific characteristics like pH, temperature, gas composition, and moisture can all be detected and measured by these tiny devices. Packaging materials that incorporate nanosensors can produce data that is helpful to supply chain management and consumers by way of visual indicators or other means (Fuertes et al., 2016).

5.5.1 OVERVIEW OF NANOSENSORS FOR REAL-TIME MONITORING

Nanosensors are intended to identify particular molecules or modifications to the physical characteristics linked to food quality. Some instances of nanosensors used to monitor the quality of food are given in detail.

5.5.1.1 pH NANOSENSORS

For monitoring and regulating the pH levels of packaged products, pH nanosensors are proving to be a viable technology in food packaging. Numerous food products' quality, safety, and shelf life are impacted by pH, which is a crucial factor. These sensors are able to gauge a food product's acidity or alkalinity. Monitoring pH fluctuations that can be an indication of spoiling or degradation is one of their main uses (Shamsipur et al., 2019). An effective way to monitor and regulate pH levels in food packaging is to use pH nanosensors. Their application can help to preserve food quality, ensure safety, and minimize food waste. To encourage their wider implementation in the food sector, ongoing research and development efforts are concentrated on improving the sensitivity, stability, and cost of pH nanosensors (Rodrigues et al 2017).

5.5.1.2 TEMPERATURE NANOSENSORS

Nanosensors that measure and regulate temperature are becoming increasingly useful in food packaging. Temperature is a crucial factor that influences the quality, safety, and shelf life of numerous food products (Fuertes et al., 2016). Food that is packaged can have its temperature monitored and detected by nanosensors. They provide real-time information on temperature changes, making it easier to make sure that the food is kept at the ideal temperature for optimum preservation (Wang et al., 2011). By providing accurate temperature monitoring and control, temperature nanosensors significantly improve the safety, quality, and shelf life of food. To make temperature nanosensors more widely employed in the food business, ongoing research and development efforts are concentrated on enhancing their sensitivity, accuracy, and affordability (Rodrigues et al., 2017).

5.5.1.3 GAS NANOSENSORS

Innovative tools that allow for the monitoring and management of gas composition within packaged food products are gas nanosensors. These nanosensors measure and detect the presence of certain gases, revealing important details about the surroundings of the food (Caon et al., 2017). Gases released by food products, such as carbon dioxide, ethylene, or volatile organic compounds, can be detected using nanosensors. Modifications in gas composition may be a sign of deterioration or the release of harmful substances (Aroutiounian, 2019). By providing accurate monitoring and control of gas composition, gas nanosensors significantly improve the quality, safety, and shelf life of food. Their sensitivity, selectivity, and affordability are the main areas of focus of ongoing research and development in order to encourage their widespread implementation in the food sector (Caon et al., 2017).

5.5.1.4 MOISTURE NANOSENSORS

Innovative devices like moisture nanosensors allow for the monitoring and regulation of moisture levels in food packaging. Moisture is a vital factor that influences the quality, texture, and shelf life of a variety of food products (Omanović-Mikličanina and Maksimović 2016). The amounts of moisture in packaged food can be determined by these sensors. They support the identification of moisture fluctuations that can cause spoilage or impair the product's texture and quality (Joyner and Kumar, 2015). Due to its ability to precisely monitor and adjust moisture levels, moisture nanosensors have a substantial impact on maintaining food quality, safety, and shelf life (Rodrigues et al., 2017).

5.5.2 INTEGRATION OF NANOSENSORS INTO PACKAGING MATERIALS

Nanosensors can be incorporated right into the packaging materials, allowing for real-time monitoring and giving customers and supply chain management useful data. Two common applications are visual indicators for consumers, data for supply chain management.

5.5.2.1 VISUAL INDICATORS FOR CONSUMERS

Visual cues or indications to notify consumers about the quality, freshness, and safety of the packed food product are crucial tools employed in food packaging. These indicators help customers make decisions since they have been designed to be simple for them to understand and interpret (Petrescu et al., 2022). Nanosensors can be engineered to respond to particular quality factors by changing color or emitting other visual signals. For instance, a color-changing nanosensor may switch from blue (fresh) to red (spoiled) to indicate the freshness of a food product (Fuertes et al., 2016). Consumers can use this visual signal to make well-informed decisions about the quality and safety of the food they eat. Food packaging visual indicators for consumers have a vital role in establishing consumer confidence, increasing transparency, and supporting with well-informed decision-making. These indicators are intended to give consumers clear and simple-to-understand information on the food product's freshness, quality, safety, sustainability, and other crucial aspects, enabling them to make decisions according to their preferences and requirements (Chelliah et al., 2021).

5.5.2.2 DATA FOR SUPPLY CHAIN MANAGEMENT

In the food business, data is vital to supply chain management. It offers insightful knowledge and data that facilitate effective decision-making, process enhancement, and improved supply chain coordination. Nanosensors can produce quality parameter data that can be remotely communicated and gathered for assessment, such as temperature, pH, or gas composition. This information can be used to monitor and maintain ideal storage conditions, spot potential supply-chain problems, and make sure that food products are handled and transported safely (Chelliah et al., 2021). In the food business, data-driven supply chain management aids in streamlining operations, cutting costs, improving food safety, raising product quality, and satisfying customer demands. Advanced analytics tools and the availability of correct data enable businesses to make wise decisions, increase supply chain visibility, and drive overall performance (Fuertes et al., 2016). The efficiency and accuracy of food quality monitoring are improved by incorporating nanosensors into packaging materials. Real-time information from these sensors on crucial quality factors enables prompt action to stop spoilage, improve storage conditions, and guarantee customer safety. Food waste can be reduced, total product quality and integrity can be enhanced, and supply chain management may be improved with the use of nanosensor-based indications and data (Tiwari et al., 2023).

5.6 SAFETY AND REGULATORY CONSIDERATIONS

For the purpose of ensuring consumer protection and addressing potential dangers, it is crucial to take into account the safety concerns and regulatory framework surrounding nanotechnology in food packaging (Tarhan, 2020). Steps in the development and application of nanotechnology-based food packaging include evaluating the safety of nanomaterials in materials that come into contact with food and adhering to regulatory standards. Here are key points to consider (Jain et al., 2018; Wesley et al., 2014).

5.6.1 SAFETY ASSESSMENT OF NANOMATERIALS IN FOOD CONTACT MATERIALS

In order to determine whether nanoparticles used in food packaging are safe, it is necessary to examine potential hazards and make sure that they are not harmful to human health. Some key factors to consider include physicochemical properties, toxicological evaluation, and migration studies.

5.6.1.1 PHYSICOCHEMICAL PROPERTIES

The physicochemical characteristics of the materials themselves are vital in assuring their efficiency and compatibility with food products when it comes to food packaging nanomaterials. Assessing the properties of nanomaterials, including size, shape, surface charge, and stability, is crucial because these characteristics might affect how they behave and how they might interact with food and the human body (Jain et al., 2018). Researchers and manufacturers may develop and optimize nanoparticles for applications in food packaging that deliver the needed functionalities while assuring the safety and quality of the packaged food products by taking into account their physicochemical qualities. Designing efficient and environmentally friendly food packaging solutions depends on accurately characterizing and comprehending these features (Wesley et al., 2014; Tarhan, 2020).

5.6.1.2 TOXICOLOGICAL EVALUATION

A crucial component of guaranteeing the safety of food packaging nanomaterials is toxicological evaluation. In order to examine the hazards connected to the use of nanomaterials in food packaging, as well as any potential toxicity and negative health impacts, this study evaluates the risks involved (Wesley et al., 2014). To assess the potential risks linked to nanomaterials, toxicological research is crucial. Examining their potential for cellular toxicity, genotoxicity, immunotoxicity, and other negative effects is part of this process. It is essential for determining the safety of food packaging nanomaterials and reducing any potential risks to human health. It guarantees that the nanoparticles used in food packaging adhere to high safety requirements, enabling the development of innovative packaging ideas that are risk-free and safeguard the safety and well-being of consumers (Jain et al., 2018; Tarhan 2020).

5.6.1.3 MIGRATION STUDIES

Studies on migration are crucial to the evaluation of the safety of food packaging materials, particularly nanomaterials. These investigations emphasize on analyzing the compounds that migrate from the packing material into the food with the goal of determining any potential dangers. In order to ascertain their possible exposure levels, it is imperative to look into the migration of nanomaterials from packaging into food (Jain et al., 2018). Migration studies analyze the impact on food safety and consumer health and determine whether nanoparticles may leach or move into the food. For evaluating the safety of food packaging materials, including nanomaterials, migration studies are essential. Potential dangers can be recognized and efficiently addressed by assessing the migration of contaminants from the container into the food, ensuring the safety of consumers and compliance with regulatory requirements (Wesley et al., 2014).

5.7 POTENTIAL APPLICATIONS OF NANOPACKAGING

Various aspects of food packaging have successfully used nanotechnology, increasing food safety, extending shelf life, and improving product quality (Ashfaq et al., 2022). Here are some particular instances and case studies demonstrating effective uses of nanotechnology in food packaging.

5.7.1 NANOCOATINGS FOR EXTENDING THE SHELF LIFE OF FRUITS

By creating a barrier of protection and regulating different physicochemical processes that lead to fruit deterioration, nanocoatings have become a potential solution for extending the shelf life of fruits. The development of fruit nanocoating to extend the shelf life of fruits is one application of nanotechnology (Ghosh and Singh, 2022).). For instance, scientists developed edible coverings using nanoparticles like zinc oxide or silver. These coatings serve as a barrier, preventing moisture loss and microbiological contamination. As a result, fruit rotting is decreased and the freshness and quality of the fruit are prolonged. Such nanocoatings have shown great results in conserving fruits such as mangoes, strawberries, and bananas. By acting as protective barriers, regulating gas exchange, controlling moisture, and combining antibacterial capabilities, nanocoatings hold considerable promise for increasing the shelf life of fruits. While preserving the quality and safety of fresh produce, ongoing research and development in this area could change fruit packaging and help reduce food waste (Correa-Pacheco et al., 2021).

5.7.2 NANOSENSORS FOR FRESHNESS MONITORING IN SEAFOOD

Nanosensors are becoming crucial tools for assessing how fresh seafood is, enabling accurate and fast evaluation of the product's quality and safety. Monitoring the freshness of seafood and ensuring its high quality and safety require nanosensors. Nanosensors that have been developed by researchers can detect biochemical changes brought on by seafood spoilage, such as the release of volatile compounds or pH changes (Manoj et al., 2021). These nanosensors, which are embedded into packaging materials, offer real-time information on the freshness of seafood products. They assist in discovering the best storage conditions and figuring out the shelf life, enabling stakeholders to make educated decisions about the caliber of the product and the rotation of the shelves. With real-time, nonintrusive, and multiparameter assessment of freshness indicators, nanosensors have substantial advantages in the freshness monitoring of seafood. Their incorporation into seafood distribution and packaging systems can help to increase quality control, cut down on waste, and guarantee the safety of seafood products, ultimately leading to an increase in customer satisfaction and confidence (Liu et al., 2022).

5.7.3 NANO-BASED ACTIVE PACKAGING FOR DAIRY PRODUCTS

As a potential strategy to extend dairy products' shelf lives, maintain freshness, and guarantee product safety, nano-based active packaging has drawn a lot of interest. To improve the preservation of dairy products, nano-based active packaging solutions have been developed. In the packaging of dairy products, for instance, antibacterial nanoparticles like silver or copper nanoparticles may be present in nanocomposite films (Mastromatteo et al., 2015). These nanoparticles produce antimicrobial substances, which slow the growth of germs that cause spoiling and increase the shelf life of dairy products. Additionally, the packaging's nanosensors can track temperature changes to make sure the products are kept in the right environments. Extended shelf life, greater product safety, and improved sensory qualities are just a few benefits of dairy products' active packaging made of nano-technology. However, it is essential to ensure the security and compliance with regulations of nanomaterials used in packaging and carry out careful analyses of their possible migration into dairy products (Aziz et al., 2022).

5.7.4 NANO-ENABLED OXYGEN BARRIER FILMS

Nano-enabled oxygen barrier films are advanced packaging materials that leverage nanotechnology to enhance their oxygen barrier properties. These films are intended to increase the shelf life of products that are sensitive to oxygen by reducing the permeability of oxygen through the packaging. Oxygen barrier films that shield food products, such as snacks, from oxidative deterioration have been developed via nanotechnology (Reščeк et al., 2016). The packaging films contain nanoparticles, such as multilayer silicate nanoclays or nanoscale metal oxides, to make oxygen molecules travel more slowly and therefore reduce their permeability. This lengthens the shelf life of packaged food goods by improving their fresh-ness and flavor stability. A significant innovation in packaging technology, nano-enabled oxygen barrier films provide better oxygen barrier qualities to safeguard and prolong the shelf life of items sensitive to oxygen (Eleftheriadou et al., 2017).

5.7.5 NANOSCALE ANTIMICROBIAL PACKAGING FOR MEAT AND POULTRY

In order to improve food safety, limit microbial development, and increase the shelf life of these perishable food items, nanoscale antimicrobial packaging for meat and poultry products has emerged as a promising alternative. To prevent microbial development and enhance food safety, beef and poultry products have been packaged with antimicrobial materials based on nanotechnology (Chawla et al., 2021). Antimicrobial nanoparticle-infused active packaging materials, including silver or zinc oxide nanoparticles, can stop bacterial contamination and increase the shelf life of these perishable goods. The danger of food spoiling and foodborne infections is decreased by these nanoparticles' release of anti-bacterial agents. The use of nanoscale antimicrobial packaging has the potential to significantly improve food safety, increase the shelf life of meat and poultry products, and ensure product quality. The effectiveness and viability of nanoscale antimicrobial packaging solutions for various foods including meat and poultry sector will be improved via continued research and development (Hettiarachchy and Ravichandran, 2012). Such case studies highlight successful applications of nanotechnology in food packaging, demonstrating its potential to enhance food preservation, maintain quality, and improve consumer safety. By leveraging nanotechnology, packaging solutions can address critical challenges in the food industry, reducing food waste and ensuring the availability of safe and high-quality food products (Sharma et al., 2017).

5.8 CHALLENGES IN NANOPACKAGING

Nanotechnology in food packaging presents both opportunities and challenges. Some challenges include *Risk Assessment:* Assessing the potential risks associated with nanomaterials requires ongoing research and the development of appropriate testing methodologies to ensure accurate and reliable results (Gao and Lowry, 2018), *Standardization:* Establishing standardized methods for the characterization and safety assessment of nanomaterials is essential to ensure consistency in evaluating their safety, *Public Perception:* Addressing public concerns about the safety and ethical implications of nanotechnology in food packaging is important for building consumer trust and acceptance (Laux et al., 2018). As the field continues to evolve, it is essential for regulatory bodies, researchers, and industry stakeholders to work collaboratively to enhance safety assessments, establish robust regulations, and ensure the responsible and sustainable use of nanotechnology in food packaging (Adeyeye, 2019).

5.9 CONCLUSION AND FUTURE PROSPECTIVES

Nanotechnology's use in food packaging has resulted in considerable advancements and encouraging results. The use of nanomaterials in numerous areas of food packaging, including active and intelligent packaging, nanocomposite films and coatings, active food preservation, nanosensors for food quality monitoring, and safety considerations, is one of the key results. By incorporating antimicrobial, antioxidant, and oxygen-scavenging properties into active packaging systems, nanotechnology has enabled the development of increasing shelf life for food products. In order to monitor food freshness, temperature, humidity, and other quality characteristics in real time, it has made it easier to incorporate nanosensors and indicators into packaging materials. Food safety is ensured by nanocomposite films and coatings because they have superior mechanical, barrier, and thermal properties that prevent microbial development and reduce the spread of harmful substances. Additionally, through the application of nanoencapsulation techniques, nanotechnology has made it possible to release active agents for food preservation in a controlled manner. The further development and commercialization of packaging solutions based on nanotechnology are dependent on continued research and collaboration. To ensure consumer safety, it is critical to evaluate the potential risks related to nanomaterials, establish standardized testing procedures, and comply with legal requirements. Understanding the long-term consequences of nanomaterials, addressing public concerns, and creating environmentally friendly and sustainable packaging alternatives using nanotechnology should be the main goals of ongoing research.

KEYWORDS

- **food packaging**
- **intelligent packaging**
- **nanocomposite films**
- **nanomaterials**
- **nanotechnology**

REFERENCES

Adegbeye, M. J., Elghandour, M. M., Barbabosa-Pliego, A., Monroy, J. C., Mellado, M., Reddy, P. R. K., & Salem, A. Z. (2019). Nanoparticles in equine nutrition: Mechanism of action and application as feed additives. *Journal of Equine Veterinary Science, 78*, 29–37.

Adeyeye, S. A. O. (2019). Food packaging and nanotechnology: safeguarding consumer health and safety. *Nutrition & Food Science, 49*(6), 1164–1179.

Ahmadi, Y., Moeini, N., Yadav, M., & Ahmad, S. (2021). Antimicrobial polymer nanocomposite films and coatings. In *Handbook of polymer nanocomposites for industrial applications* (pp. 379–397). Elsevier.

Aroutiounian, V. M. (2019). Gas nanosensors made from semiconductor metal oxides. *Journal of Contemporary Physics (Armenian Academy of Sciences), 54*, 356–367.

Ashfaq, A., Khursheed, N., Fatima, S., Anjum, Z., & Younis, K. (2022). Application of nanotechnology in food packaging: Pros and Cons. *Journal of Agriculture and Food Research, 7*, 100270.

Ayala-Fuentes, J. C., & Chavez-Santoscoy, R. A. (2021). Nanotechnology as a Key to Enhance the Benefits and Improve the Bioavailability of Flavonoids in the Food Industry. *Foods, 10*(11), 2701.

Aziz, S. A. A. A., Mahmoud, R., & Mohamed, M. B. E. D. (2022). Control of biofilm-producing Pseudomonas aeruginosa isolated from dairy farm using Virokill silver nano-based disinfectant as an alternative approach. *Scientific Reports, 12*(1), 9452.

Bajpai, V. K., Kamle, M., Shukla, S., Mahato, D. K., Chandra, P., Hwang, S. K., ... & Han, Y. K. (2018). Prospects of using nanotechnology for food preservation, safety, and security. *Journal of Food and Drug Analysis, 26*(4), 1201–1214.

Basumatary, I. B., Mukherjee, A., Katiyar, V., & Kumar, S. (2022). Biopolymer-based nanocomposite films and coatings: recent advances in shelf-life improvement of fruits and vegetables. *Critical Reviews in Food Science and Nutrition, 62*(7), 1912–1935.

Beshai, H., Sarabha, G. K., Rathi, P., Alam, A. U., & Deen, M. J. (2020). Freshness monitoring of packaged vegetables. *Applied Sciences, 10*(21), 7937.

Bratovčić, A., Odobašić, A., Ćatić, S., & Šestan, I. (2015). Application of polymer nanocomposite materials in food packaging. *Croatian journal of food science and technology, 7*(2), 86–94.

Bumbudsanpharoke, N., & Ko, S. (2019). Nanomaterial-based optical indicators: Promise, opportunities, and challenges in the development of colorimetric systems for intelligent packaging. *Nano Research, 12*, 489–500.

Caon, T., Martelli, S. M., & Fakhouri, F. M. (2017). New trends in the food industry: application of nanosensors in food packaging. In *Nanobiosensors* (pp. 773–804). Academic Press.

Chausali, N., Saxena, J., & Prasad, R. (2022). Recent trends in nanotechnology applications of bio-based packaging. *Journal of Agriculture and Food Research, 7*, 100257.

Chawla, R., Sivakumar, S., & Kaur, H. (2021). Antimicrobial edible films in food packaging: Current scenario and recent nanotechnological advancements-a review. *Carbohydrate Polymer Technologies and Applications, 2*, 100024.

Chelliah, R., Wei, S., Daliri, E. B. M., Rubab, M., Elahi, F., Yeon, S. J., ... & Oh, D. H. (2021). Development of nanosensors based intelligent packaging systems: food quality and medicine. *Nanomaterials, 11*(6), 1515.

Correa-Pacheco, Z. N., Corona-Rangel, M. L., Bautista-Baños, S., & Ventura-Aguilar, R. I. (2021). Application of natural-based nanocoatings for extending the shelf life of green bell pepper fruit. *Journal of Food Science, 86*(1), 95–102.

Dhar, P., Tarafder, D., Kumar, A., & Katiyar, V. (2015). Effect of cellulose nanocrystal polymorphs on mechanical, barrier and thermal properties of poly (lactic acid) based bionanocomposites. *RSC Advances, 5*(74), 60426–60440.

Dodero, A., Escher, A., Bertucci, S., Castellano, M., & Lova, P. (2021). Intelligent packaging for real-time monitoring of food-quality: Current and future developments. *Applied Sciences, 11*(8), 3532.

Eleftheriadou, M., Pyrgiotakis, G., & Demokritou, P. (2017). Nanotechnology to the rescue: using nano-enabled approaches in microbiological food safety and quality. *Current Opinion in Biotechnology, 44*, 87–93.

Enescu, D., Cerqueira, M. A., Fucinos, P., & Pastrana, L. M. (2019). Recent advances and challenges on applications of nanotechnology in food packaging. A literature review. *Food and Chemical Toxicology, 134*, 110814.

Fuertes, G., Soto, I., Carrasco, R., Vargas, M., Sabattin, J., & Lagos, C. (2016). Intelligent packaging systems: sensors and nanosensors to monitor food quality and safety. *Journal of Sensors, 2016*.

Gao, X., & Lowry, G. V. (2018). Progress towards standardized and validated characterizations for measuring physicochemical properties of manufactured nanomaterials relevant to nano health and safety risks. *NanoImpact, 9*, 14–30.

Ghosh, M., & Singh, A. K. (2022). Potential of engineered nanostructured biopolymer based coatings for perishable fruits with Coronavirus safety perspectives. *Progress in Organic Coatings, 163*, 106632.

Han, W., Yu, Y., Li, N., & Wang, L. (2011). Application and safety assessment for nano-composite materials in food packaging. *Chinese Science Bulletin, 56*(12), 1216–1225.

Hernández-Díaz, J. A., Garza-García, J. J., Zamudio-Ojeda, A., León-Morales, J. M., López-Velázquez, J. C., & García-Morales, S. (2021). Plant-mediated synthesis of nanoparticles and their antimicrobial activity against phytopathogens. *Journal of the Science of Food and Agriculture, 101*(4), 1270–1287.

Hettiarachchy, N., & Ravichandran, M. (2012). Potential use of edible nanoscale coatings for meat. *Food Safety Magazine,* 4–5.

Horev, B., Sela, S., Vinokur, Y., Gorbatsevich, E., Pinto, R., & Rodov, V. (2012). The effects of active and passive modified atmosphere packaging on the survival of Salmonella enterica serotype Typhimurium on washed romaine lettuce leaves. *Food Research International, 45*(2), 1129–1132.

Imran, M., Revol-Junelles, A. M., Martyn, A., Tehrany, E. A., Jacquot, M., Linder, M., & Desobry, S. (2010). Active food packaging evolution: transformation from micro-to nanotechnology. *Critical Reviews in Food Science and Nutrition, 50*(9), 799–821.

Jafarzadeh, S., & Jafari, S. M. (2021). Impact of metal nanoparticles on the mechanical, barrier, optical and thermal properties of biodegradable food packaging materials. *Critical Reviews in Food Science and Nutrition, 61*(16), 2640–2658.

Jain, A., Ranjan, S., Dasgupta, N., & Ramalingam, C. (2018). Nanomaterials in food and agriculture: an overview on their safety concerns and regulatory issues. *Critical Reviews in Food Science and Nutrition, 58*(2), 297–317.
Joyner, J. J., & Kumar, D. V. (2015). Nanosensors and their applications in food analysis: a review. *The International Journal of Science and Technoledge, 3*(4), 80.
Khaneghah, A. M., Hashemi, S. M. B., & Limbo, S. (2018). Antimicrobial agents and packaging systems in antimicrobial active food packaging: An overview of approaches and interactions. *Food and Bioproducts Processing, 111*, 1–19.
Kumar, S., Mukherjee, A., & Dutta, J. (2020). Chitosan based nanocomposite films and coatings: Emerging antimicrobial food packaging alternatives. *Trends in Food Science & Technology, 97*, 196–209.
Kuswandi, B. (2016). Nanotechnology in food packaging. *Nanoscience in Food and Agriculture, 1*, 151–183.
Laux, P., Tentschert, J., Riebeling, C., Braeuning, A., Creutzenberg, O., Epp, A., ... & Luch, A. (2018). Nanomaterials: certain aspects of application, risk assessment and risk communication. *Archives of Toxicology, 92*, 121–141.
Liu, Y., Xiao, Y., Shang, M., Zhuang, Y., & Wang, L. (2022). Smart fluorescent tag based on amine response for non-contact and visual monitoring of seafood freshness. *Chemical Engineering Journal, 428*, 132647.
Manoj, D., Shanmugasundaram, S., & Anandharamakrishnan, C. (2021). Nanosensing and nanobiosensing: Concepts, methods, and applications for quality evaluation of liquid foods. *Food Control, 126*, 108017.
Manzoor, A., Khan, S., Dar, A. H., Pandey, V. K., Shams, R., Ahmad, S., & Pandiselvam, R. (2023). Recent insights into green antimicrobial packaging towards food safety reinforcement: A review. *Journal of Food Safety*, e13046.
Mastromatteo, M., Conte, A., Lucera, A., Saccotelli, M. A., Buonocore, G. G., Zambrini, A. V., & Del Nobile, M. A. (2015). Packaging solutions to prolong the shelf life of Fiordilatte cheese: Bio-based nanocomposite coating and modified atmosphere packaging. *LWT-Food Science and Technology, 60*(1), 230–237.
Mlalila, N., Kadam, D. M., Swai, H., & Hilonga, A. (2016). Transformation of food packaging from passive to innovative via nanotechnology: concepts and critiques. *Journal of Food Science and Technology, 53*, 3395–3407.
Mohammadpour, Z., & Naghib, S. M. (2021). Smart nanosensors for intelligent packaging. In *Nanosensors for Smart Manufacturing* (pp. 323–346). Elsevier.
Neethirajan, S., & Jayas, D. S. (2011). Nanotechnology for the food and bioprocessing industries. *Food and Bioprocess Technology, 4*, 39–47.
Nikolic, M. V., Vasiljevic, Z. Z., Auger, S., & Vidic, J. (2021). Metal oxide nanoparticles for safe active and intelligent food packaging. *Trends in Food Science & Technology, 116*, 655–668.
Omanović-Mikličanina, E., & Maksimović, M. (2016). Nanosensors applications in agriculture and food industry. *Bull Chem Technol Bosnia Herzegovina, 47*, 59–70.
Pavelková, A. (2013). Time temperature indicators as devices intelligent packaging. *Acta Universitatis Agriculturae et Silviculturae Mendelianae Brunensis, 61*(1), 245–251.
Pereira de Abreu, D. A., Cruz, J. M., & Paseiro Losada, P. (2012). Active and intelligent packaging for the food industry. *Food Reviews International, 28*(2), 146–187.

Petrescu, D. C., Vermeir, I., Burny, P., & Petrescu-Mag, R. M. (2022). Consumer evaluation of food quality and the role of environmental cues. A comprehensive cross-country study. *European Research on Management and Business Economics, 28*(2), 100178.
Primožič, M., Knez, Ž., & Leitgeb, M. (2021). (Bio) Nanotechnology in food science—Food packaging. *Nanomaterials, 11*(2), 292.
Rai, M., Ingle, A. P., Gupta, I., Pandit, R., Paralikar, P., Gade, A., ... & dos Santos, C. A. (2019). Smart nanopackaging for the enhancement of food shelf life. *Environmental Chemistry Letters, 17*, 277–290.
Reščék, A., Ščetar, M., Hrnjak-Murgić, Z., Dimitrov, N., & Galić, K. (2016). Polyethylene/polycaprolactone nanocomposite films for food packaging modified with magnetite and casein: oxygen barrier, mechanical, and thermal properties. *Polymer-Plastics Technology and Engineering, 55*(14), 1450–1459.
Rodrigues, S. M., Demokritou, P., Dokoozlian, N., Hendren, C. O., Karn, B., Mauter, M. S., & Lowry, G. V. (2017). Nanotechnology for sustainable food production: promising opportunities and scientific challenges. *Environmental Science: Nano, 4*(4), 767–781.
Sen, M. (2020). Nanocomposite materials. *Nanotechnology and the Environment,* 1–12.
Shamsipur, M., Barati, A., & Nematifar, Z. (2019). Fluorescent pH nanosensors: Design strategies and applications. *Journal of Photochemistry and Photobiology C: Photochemistry Reviews, 39*, 76–141.
Sharma, C., Dhiman, R., Rokana, N., & Panwar, H. (2017). Nanotechnology: An untapped resource for food packaging. *Frontiers in Microbiology, 8*, 1735.
Silvestre, C., Duraccio, D., & Cimmino, S. (2011). Food packaging based on polymer nanomaterials. *Progress in Polymer Science, 36*(12), 1766–1782.
Singh, P. (2018). Nanotechnology in food preservation. *Food Science, 9*, 435–441.
Siró, I. (2012). Active and intelligent packaging of food. *Progress in Food Preservation,* 23–48.
Srividya, N., Ghoora, M. D., & Padmanabh, P. R. (2017). Antimicrobial nanotechnology: Research implications and prospects in food safety. In *Food preservation* (pp. 125–165). Academic Press.
Tarhan, Ö. (2020). Safety and regulatory issues of nanomaterials in foods. *Handbook of Food Nanotechnology,* 655–703.
Temgire, S., Borah, A., Kumthekar, S., & Idate, A. (2021). Recent trends in ready to eat/cook food products. *The Pharma Innovation Journal, 10*(5), 211–217.
Thiruvengadam, M., Rajakumar, G., & Chung, I. M. (2018). Nanotechnology: current uses and future applications in the food industry. *3 Biotech, 8*, 1–13.
Tiwari, A., Barbhuiya, R. I., & Dash, K. K. (2023). Nanotechnology-based sensors for shelf-life determination of food materials. In *Nanotechnology applications for food safety and quality monitoring* (pp. 289–300). Academic Press.
Unalan, I. U., Cerri, G., Marcuzzo, E., Cozzolino, C. A., & Farris, S. (2014). Nanocomposite films and coatings using inorganic nanobuilding blocks (NBB): current applications and future opportunities in the food packaging sector. *RSC Advances, 4*(56), 29393–29428.
Venkatesan, R., Rajeswari, N., & Tamilselvi, A. (2018). Antimicrobial, mechanical, barrier, and thermal properties of bio-based poly (butylene adipate-co-terephthalate) (PBAT)/Ag_2 O nanocomposite films for packaging application. *Polymers for Advanced Technologies, 29*(1), 61–68.

Vieira, I. R. S., de Carvalho, A. P. A. D., & Conte-Junior, C. A. (2022). Recent advances in biobased and biodegradable polymer nanocomposites, nanoparticles, and natural antioxidants for antibacterial and antioxidant food packaging applications. *Comprehensive Reviews in Food Science and Food Safety, 21*(4), 3673–3716.

Wang, S., Liu, X., Yang, M., Zhang, Y., Xiang, K., & Tang, R. (2015). Review of time temperature indicators as quality monitors in food packaging. *Packaging Technology and Science, 28*(10), 839–867.

Wang, X. D., Song, X. H., He, C. Y., Yang, C. J., Chen, G., & Chen, X. (2011). Preparation of reversible colorimetric temperature nanosensors and their application in quantitative two-dimensional thermo-imaging. *Analytical Chemistry, 83*(7), 2434–2437.

Wesley, S. J., Raja, P., Raj, A. A., & Tiroutchelvamae, D. (2014). Review on-nanotechnology applications in food packaging and safety. *International Journal of Engineering Research, 3*(11), 645–651.

Wilson, M. D., Stanley, R. A., Eyles, A., & Ross, T. (2019). Innovative processes and technologies for modified atmosphere packaging of fresh and fresh-cut fruits and vegetables. *Critical Reviews in Food Science and Nutrition, 59*(3), 411–422.

Wu, D., Zhang, M., Chen, H., & Bhandari, B. (2021). Freshness monitoring technology of fish products in intelligent packaging. *Critical Reviews in Food Science and Nutrition, 61*(8), 1279–1292.

Yildirim, S., & Röcker, B. (2018). Active packaging. In *Nanomaterials for food packaging* (pp. 173–202). Elsevier.

Yousefi, H., Su, H. M., Imani, S. M., Alkhaldi, K., M. Filipe, C. D., & Didar, T. F. (2019). Intelligent food packaging: A review of smart sensing technologies for monitoring food quality. *ACS Sensors, 4*(4), 808–821.

Youssef, A. M., Assem, F. M., El-Sayed, S. M., Salama, H., & Abd El-Salam, M. H. (2017). Utilization of edible films and coatings as packaging materials for preservation of cheeses. *Journal of Packaging Technology and Research, 1*, 87–99.

Zhang, S., & Ali, N. (Eds.). (2007). *Nanocomposite thin films and coatings: Processing, properties and performance*. Imperial College Press.

CHAPTER 6

Food Bioremediation Using Nanotechnology

RAFEEYA SHAMS[1,*], VINAY KUMAR PANDEY[2], JAGMOHAN SINGH[3], KSHIROD KUMAR DASH[4], and AAMIR HUSSAIN DAR[5]

[1]*Department of Food Technology and Nutrition, Lovely Professional University, Phagwara, Punjab, India*

[2]*Research and Development Cell, Biotechnology Department, Manav Rachna International Institute of Research and Studies (Deemed to Be University) Faridabad, Haryana, India*

[3]*Division of Food Science and Technology, Sher-e-Kashmir University of Agricultural Sciences and Technology, Jammu, India*

[4]*Department of Food Processing Technology, Ghani Khan Choudhury Institute of Engineering and Technology, Malda, West Bengal, India*

[5]*Department of Food Technology, Islamic University of Science and Technology, Awantipora, Kashmir, India*

*Corresponding author

ABSTRACT

Food bioremediation, the use of biological agents to remove or reduce pollutants from food items, is a developing field that has important implications for environmental sustainability and food safety. With its distinctive qualities and broad range of uses, nanotechnology presents intriguing options for improving food bioremediation procedures.

Nanotechnological Advancements in the Food Industry.
Ishrat Majid, Bababode Adesegun Kehinde, and Sheikh Shafat Hussain (Eds.)

DOI: 10.1201/9781774919750-6

Food bioremediation based on nanotechnology is crucial because it offers inventive ways to remove contaminants from food products. Pollutants are identified, dismantled, and purified utilizing nanomaterials and nanodevices to enhance food safety and quality. This process uses biological agents and nanotechnology to get rid of food-borne contamination, making food manufacturing safer and more environmentally friendly. This chapter seeks to provide a thorough review of the application of nanotechnology in food bioremediation by examining the fundamentals, difficulties, possibilities, and long-term outlooks in this area.

6.1 INTRODUCTION

For the cleanup and repair of contaminated environments, bioremediation is a practical and economical solution. It entails the use of biological agents to break down, change, or eliminate contaminants from soil, water, or the air. These agents can be bacteria, plants, or enzymes. The inherent capabilities of living things to degrade or transform toxic compounds into less dangerous forms are used in this natural remediation process to lessen the impact of toxins on the environment (Azubuike et al., 2016). It is impossible to exaggerate the role that bioremediation plays in environmental sustainability. Traditional remediation techniques like excavation and disposal, incineration, or chemical treatments can have downsides like high costs, environmental damage, and additional waste production. On the other hand, bioremediation has a number of benefits that make it an attractive and long-term approach to environmental cleanup. First off, bioremediation is a harmless, natural process. Ecosystems can be preserved and restored because it uses nature's cleansing capacity to remove contaminants (Kumar et al., 2018). This strategy encourages the restoration of natural habitats while causing the least amount of environmental disruption. Second, bioremediation is an economical technique. It can be more affordable than conventional remediation methods since it uses less energy, less material and equipment, and less transportation. Furthermore, bioremediation is frequently possible on-site, negating the need for pricey off-site disposal. Additionally, bioremediation is a sustainable strategy that supports the ideas of resource conservation and the circular economy. Closing the loop and lessening the total impact on the environment, it enables the conversion of pollutants into benign compounds rather than merely moving toxins from one place to another. Additionally, bioremediation is flexible and adaptable (Alori et al., 2022). Specific microbes or enzymatic systems can target different types of contaminants, allowing for customized solutions based on the contaminant and the site features. Numerous contaminants, such as petroleum hydrocarbons, heavy metals, pesticides, and organic pollutants, can be treated via bioremediation. Furthermore, bioremediation has the capacity to clean up contaminated locations that would otherwise be challenging to handle. To increase total cleanup effectiveness, it can be utilized in concert with other remediation strategies including physical or chemical procedures. In order to get the best results, this integrated strategy—hybrid remediation—combines the advantages of many approaches (Bala et al., 2022).

The control and manipulation of matter at the nanoscale, which is typically between 1 and 100 nm, is the focus of the multidisciplinary area of nanotechnology. At this scale, materials display distinctive physical, chemical, and biological characteristics that can be used for a variety of purposes in a variety of industries. Nanotechnology has a wide range of uses that have the potential to transform businesses and have a significant impact on many facets of our lives. The fields of electronics and computing are one well-known area where nanotechnology has made substantial advancements (Bayda et al., 2019). High-performance computing, data storage, and energy-efficient gadgets have all advanced as a result of the use of nanoscale materials like carbon nanotubes and nanowires to make quicker and smaller electronic components. Innovative uses in healthcare, communication, and consumer electronics are now possible because of the development of flexible and wearable electronics made possible by nanotechnology. Nanotechnology has shown a lot of potential in the medical and healthcare industries. Treatment delivery to particular bodily sites using nanoparticles can increase treatment efficacy and minimize negative effects. The application of these nanoparticles in medical imaging enables earlier disease identification and more accurate disease diagnosis (Song et al., 2022).

Additionally, nanotechnology has aided in the creation of biosensors and diagnostic tools that allow for the quick and accurate identification of pathogens and biomarkers. In tissue engineering, nanomaterials have been utilized to build scaffolds for regenerative medicine that help injured tissues grow and heal. The field of energy has also benefited from the use of nanotechnology. As a way to collect a wider spectrum of light and improve energy conversion, nanomaterials like quantum dots and nanocrystals have been employed to increase the efficiency of solar cells (Shi et al., 2010). With the advent of nanoscale materials for batteries and supercapacitors, which enable better energy densities and quicker charging periods, nanotechnology is also being investigated for energy storage. Furthermore, the development of materials and technologies for the creation and storage of sustainable energy is greatly aided by nanotechnology. Nanotechnology has also aided the study of sustainability and the environment. Nanomaterials have been used to remove impurities and improve water quality during wastewater treatment and water purification (Chen and Dai, 2013). For environmental monitoring, nanosensors and nanodevices have been created, allowing for the real-time detection and measurement of contaminants. Nanotechnology has also been used to create green and sustainable materials, clean up contaminated locations, and purify the air. Nanotechnology has created new opportunities for the design and production of innovative materials with improved qualities in the field of materials science (Javaid et al., 2021). Nanocomposites, made of nanoparticles encapsulated in a matrix, have better electrical conductivity, strength, and durability. Nanocoatings offer improved defence against wear, fouling, and corrosion. The exact control of material properties at the nanoscale made possible by nanofabrication techniques has led to the creation of materials with tailored properties for certain purposes (Farooq et al., 2022).

A new field called food bioremediation uses nanotechnology to address problems with food contamination by fusing the principles of bioremediation with the special qualities of nanomaterials. Food products that have been contaminated with different pollutants pose serious threats to both human health and the environment. Chemical treatments are frequently used in traditional methods of ensuring the safety and quality of food; however, these techniques might have drawbacks and produce unwanted byproducts (Bala et al., 2022). A possible alternative strategy to effectively and sustainably reduce foodborne toxins is food bioremediation utilizing nanotechnology. Using biological agents, such as bacteria or enzymes, to degrade, transform, or eliminate pollutants from the environment is the idea behind bioremediation. It makes use of these biological agents' innate capacities to transform dangerous chemicals into less damaging ones. Although bioremediation has been effectively used in the cleanup of soil and water, interest in its use in the food business is growing. Contrarily, nanotechnology works with objects and materials that are typically between 1 and 100 nm in size (Pramanik et al., 2021). Materials exhibit distinct physical, chemical, and biological characteristics at this scale that can be used for a variety of purposes. Nanomaterials are the best options for tackling the problems associated with food contamination because of their high reactivity, large surface area-to-volume ratios, and improved transport qualities. The use of nanomaterials in food bioremediation entails maximizing the effectiveness and efficiency of bioremediation procedures in the food sector. In order to increase the general safety and quality of food products, specific pollutants can be detected, degraded, or removed using nanomaterials. They provide targeted and regulated ways to address many types of foodborne contaminants by acting as catalysts, transporters, or sensors (Ningthoujam et al., 2022).

The identification and monitoring of food pollutants is one of the main uses of nanotechnology in food bioremediation. To identify and measure the presence of different pollutants in food products, such as pathogens, poisons, or chemical residues, nanoscale biosensors can be created. These biosensors provide sensitive and quick detection capabilities, allowing for quick responses and guaranteeing the security of food supplies. According to Kumar et al. (2020), nanotechnology is essential for the deterioration and elimination of food pollutants. As catalysts, nanoparticles can quicken the conversion of contaminants into nontoxic compounds. The effectiveness and speed of pollutant removal can be increased by enhancing the activity of enzymes or microbial systems involved in biodegradation processes. To ensure targeted and regulated remediation processes, nanocarriers can be used to transport bioactive substances, such as enzymes or advantageous microbes, to specific food matrices. Additionally, advanced filtration and purification methods for food processing are provided by nanotechnology. The quality and safety of the finished food products can be improved by using nanofiltration membranes and nanoscale filtration systems to remove pollutants, particles, or undesired components from food items (Alshawwa et al., 2022).

6.2 PRINCIPLES OF BIOREMEDIATION

The use of biological techniques and agents to eliminate or degrade pollutants from food systems is known as food bioremediation. It entails the remediation or detoxification of food products that are tainted with various contaminants or unwanted compounds by the use of bacteria, enzymes, or other biological agents. By successfully removing or lowering the amounts of harmful pollutants in food, food bioremediation aims to increase food safety, quality, and sustainability (Abatenh et al., 2017).

6.2.1 CONTAMINANT REMOVAL

The removal or reduction of pollutants found in food items is the main objective of food bioremediation. Chemical pollutants, heavy metals, pesticides, mycotoxins, antibiotics, and other dangerous compounds are examples of contaminants. In order to make the food safe for consumption, it is important to use bioremediation techniques that precisely target and breakdown these toxins (Goutam et al., 2022).

6.2.2 DEGRADATION OF UNDESIRABLE COMPOUNDS

The goal of food bioremediation is to convert or destroy unwanted substances that are present in food products. Reduced levels of naturally occurring toxins, such as aflatoxins and cyanotoxins, which can be created by specific molds and algae, are one aspect of this. These substances are transformed into less hazardous or nontoxic forms using bioremediation agents such as enzymes or microbes (Edwards and Lawton, 2009).

6.2.3 PRESERVATION OF NUTRITIONAL VALUE

During the remediation process, food bioremediation aims to maintain the nutritional value of food products. Contrary to some chemical or physical decontamination techniques, bioremediation procedures can selectively target pollutants while minimizing the effects on the food's nutritive value and sensory qualities. This guarantees that the food that has been remedied maintains its quality and nutritional content (Li et al., 2022).

6.2.4 ENVIRONMENTAL SUSTAINABILITY

By offering an environmentally benign substitute for conventional decontamination techniques, food bioremediation supports environmental sustainability. Utilizing biological agents and methods lessens the need for chemical treatments, which have a negative impact on the environment. By utilizing the inherent capacity of microorganisms or enzymes to break down pollutants, bioremediation techniques can be more environmentally friendly and sustainable (Kumar et al., 2018).

6.2.5 COST-EFFECTIVENESS

The goal of food bioremediation is to provide a more affordable means of enhancing food quality and safety. In comparison to traditional remediation techniques, it can provide a more accessible and economical solution by relying on natural biological processes. It is a viable solution for both large and small food production systems because bioremediation agents may be manufactured on a large scale and the methods are simple to scale up (Murillo-Tovar et al., 2021).

6.3 NANOTECHNOLOGY IN BIOREMEDIATION

6.3.1 NANOMATERIALS FOR ENVIRONMENTAL POLLUTANT DETECTION AND MONITORING

The field of environmental contaminant detection and monitoring has undergone a revolution thanks to nanomaterials' improved sensitivity, selectivity, and adaptability. These materials, which are typically nanoscale in size, have special physicochemical characteristics that make them very efficient at sensing and detecting many types of environmental pollutants (Su et al., 2012). To specifically bind to or interact with target contaminants, nanomaterials can be functionalized with particular receptors, probes, or ligands. With the help of this functionalization, contaminants can be identified and measured very precisely. Nanomaterials can be functionalized with antibodies, aptamers, or molecular imprinting to produce sensor platforms with a strong affinity for certain contaminants. Nanomaterials with special optical characteristics, such as fluorescence, luminescence, or plasmonic effects, include quantum dots, metallic nanoparticles, and nanoclusters. These characteristics can be used to monitor and detect contaminants. A measurable signal that can be used for qualitative or quantitative analysis is produced by changes in the optical signals that nanomaterials emit or absorb when contaminants are present (Fu and Xiang, 2020). Nanomaterials with remarkable electrical conductivity and sensitivity include carbon nanotubes, graphene, and conducting polymers. These nanoparticles can be used in sensor systems to detect changes in electrical characteristics, such as resistance or conductivity, when contaminants come into contact with the nanomaterial surface. For the purpose of monitoring pollutants, this electrical reaction serves as a detecting signal. Based on their electrochemical characteristics, nanomaterials can be used in electrochemical sensors to identify and measure contaminants (Mohamed and Awad, 2022). In electrochemical sensor systems, nanomaterials such as metal nanoparticles, metal oxides, or carbon-based nanomaterials can act as electrode modifiers or catalysts. Pollutant concentration can be determined and correlated using changes in current, potential, or impedance that are caused by the interaction of contaminants with nanomaterials at the electrode surface. The unique magnetic properties of magnetic nanomaterials, such as magnetic nanoparticles, can be used to identify pollutants. The ability to selectively bind to contaminants and segregate them using magnetic fields is achieved by functionalizing nanomaterials with certain ligands or receptors (Curulli, 2020).

The concentration of pollutants in the sample can be determined by quantifying the nanomaterials that were captured. To detect and monitor gaseous contaminants, gas sensors can make use of nanomaterials like metal oxides, nanowires, or nanotubes. Electrical, optical, or thermal properties change as a result of the interaction between gas molecules and the surface of the nanomaterial, and these changes can be evaluated to establish the presence and level of pollutants (Kong et al., 2021). For on-site or in-field pollutant monitoring, portable and field-deployable systems can use nanomaterial-based sensors. Nanomaterial sensors are well suited for in-the-moment, real-time analysis due to their small size, low power needs, and compatibility with microelectronics. These tools make it possible to quickly and conveniently monitor contaminants in diverse environmental contexts. Comprehensive pollution monitoring is made possible by the design flexibility of nanomaterial-based sensors, which may measure numerous parameters simultaneously. Multiple contaminants or environmental factors can be detected and quantified concurrently by combining various nanomaterials or sensor platforms into a single device, offering a comprehensive picture of the environmental conditions (Arduini et al., 2020).

6.3.2 NANOSENSORS FOR REAL-TIME MONITORING OF BIOREMEDIATION PROCESSES

As effective tools for real-time monitoring of bioremediation processes, nanosensors have recently come to the fore. These sensors make use of nanomaterials with special qualities including high sensitivity, selectivity, and responsiveness to identify and measure numerous bioremediation-related factors. Real-time monitoring enhances the efficiency and effectiveness of bioremediation processes by enabling prompt intervention and optimization (Mahmoud and Fawzy, 2021).

6.3.2.1 DETECTION OF CONTAMINANTS

Nanosensors provide real-time pollutant detection and quantification during bioremediation procedures. They use particular nanomaterials—nanoparticles or nanotubes, for example—that have been functionalized with receptors or probes that have affinity or selectivity for the desired pollutants. A detectable signal that can be recognized and measured is produced when the contaminants interact with the nanomaterials, changing their optical, electrical, or magnetic properties (Gaviria-Arroyave et al., 2020).

6.3.2.2 MONITORING OF MICROBIAL ACTIVITY

The monitoring of microbial activity during bioremediation is made possible using nanosensors. They are able to identify microorganisms that are present and actively converting or degrading pollutants. Nanosensors offer insights into the efficacy and development of bioremediation processes by monitoring parameters including microbial growth, enzyme activity, or metabolic byproducts (Karigar and Rao, 2011).

6.3.2.3 MEASUREMENT OF ENVIRONMENTAL PARAMETERS

Nanosensors are capable of measuring a wide range of environmental factors that are important to bioremediation, including pH, temperature, oxygen saturation, and nutrient concentrations. These variables affect how well and effectively bioremediation techniques work. In order to improve the activity and growth of bioremediation agents, these parameters can be adjusted and optimized in real time in response to environmental circumstances (Bahrulolum et al., 2021).

6.3.2.4 QUANTIFICATION OF DEGRADATION PRODUCTS

The breakdown byproducts or intermediates created during bioremediation can be measured by nanosensors. Nanosensors can measure and quantify these compounds to offer data on the effectiveness and progress of pollutant reduction. This enables the evaluation of the bioremediation process's completion as well as the detection of any potential byproducts or metabolites that may need additional monitoring (Chopra and Kumar, 2020).

6.3.2.5 REMOTE SENSING AND WIRELESS COMMUNICATION

Nanosensors can be created to support wireless communication and remote sensing. This function enables *in situ* or difficult-to-reach bioremediation processes to be monitored in real time. In order to monitor and manage the bioremediation process continuously, wireless connectivity enables data to be transmitted to a central monitoring system (Vaneev et al., 2022).

6.3.2.6 MULTIPARAMETER MONITORING

Multiple parameters can be measured simultaneously by nanosensors. By collecting a comprehensive picture of the system's functioning, this capacity enables thorough monitoring of bioremediation processes. The detection of complicated interactions and the provision of a more precise assessment of the bioremediation progress are both made possible by the simultaneous monitoring of various parameters (Singh and Saxena, 2022).

6.3.2.7 MINIATURIZATION AND INTEGRATION

Nanosensors can be field-deployable platforms or portable devices that have been miniaturized. They are therefore suited for monitoring bioremediation activities on-site or in the field. Real-time monitoring in various environmental circumstances is made possible by miniaturization's simple handling and interaction with other technologies (Willner and Vikesland, 2018).

6.3.2.8 REAL-TIME FEEDBACK AND CONTROL

Nanosensor-based real-time monitoring offers feedback and control mechanisms to enhance bioremediation procedures. Nanosensors provide real-time decision-making and modifications by providing rapid data on essential parameters. The efficiency and effectiveness of the bioremediation process are increased because of this feedback loop (Mohamed and Awad, 2022).

6.4 NANOTECHNOLOGY APPROACHES FOR FOOD BIOREMEDIATION

6.4.1 NANOSCALE BIOSENSORS FOR RAPID DETECTION OF FOOD CONTAMINANTS

A potent new technique for the quick and accurate detection of food pollutants is the nanoscale biosensor. Human health is seriously endangered when food products are contaminated with diseases, poisons, chemical residues, and allergies. Traditional techniques of food testing and analysis can be costly, time-consuming, and equipment-intensive. By providing quick, sensitive, and portable detection platforms for food safety applications, nanoscale biosensors offer a viable way to get around these constraints (Wang and Duncan, 2017). Nanoscale biosensors are made up of nanomaterials like nanoparticles or nanowires that have been functionalized with biological recognition components like antibodies, enzymes, or DNA probes. Target analytes can be detected effectively and sensitively thanks to the special characteristics of nanomaterials, such as their large surface area-to-volume ratio and improved transport capabilities (Naresh and Lee, 2021). The high sensitivity of nanoscale biosensors is one of their main advantages. Due to the sensing components' efficient interaction with the target analytes, which is made possible by their nanoscale size, signal amplification and detection limits are improved. Due to its sensitivity, it is possible to detect pollutants in food samples that may not be seen using more traditional techniques. Faster detection times are made possible by the nanomaterials' tiny size and large surface area, which promote rapid diffusion and interaction with target analytes. For real-time monitoring and quality control in food processing and delivery (Willner and Vikesland, 2018), a quick response is essential. The biological recognition components that are integrated into the nanomaterials can be specifically engineered to target specific pollutants, providing exceptional selectivity and specificity for nanoscale biosensors. Due to the lack of interference from other chemicals, this specificity guarantees that the biosensors can distinguish between the presence of particular infections, poisons, or chemical residues in food samples.

Additional benefits of nanoscale biosensors include their portability and simplicity of usage. On-site testing and point-of-care applications are made possible by their compact size and compatibility with portable gadgets like smartphones or handheld readers. From farm to fork, this portability enables quick screening of food products at multiple points along the supply chain, enabling prompt responses and assuring food safety. The creation of nanoscale biosensors for the detection of food contamination has made use of a variety of nanomaterials. Because of their distinctive optical characteristics, gold nanoparticles, for instance, have been frequently used in colorimetric or plasmonic detection techniques (Seo et al., 2019). Semiconductor nanoparticles called quantum dots have fluorescence features that allow for the sensitive and multiplexed detection of many pollutants. Excellent electrical conductivity is provided by carbon nanotubes and nanowires, which can be used into electronic sensing devices to enable label-free and real-time monitoring (Khazaei et al., 2023). Nanoscale biosensors have a wide range of uses for the quick detection of food pollutants. They have been used to identify bacterial pathogens in food samples, including *Salmonella*, *Escherichia coli*, and *Listeria monocytogenes*. These biosensors can also pick up mycotoxins produced by fungi or bacteria like *Staphylococcus aureus* enterotoxin. In addition, pesticide, antibiotic, and heavy metal residues, as well as allergenic proteins in food products, have been detected using nanoscale biosensors (Ali et al., 2020).

6.4.2 NANOCARRIERS FOR TARGETED DELIVERY OF BIOACTIVE AGENTS

For the targeted delivery of bioactive chemicals in a variety of industries, including medicine, agriculture, and the food sector, nanocarriers have shown promise. The therapeutic efficacy, side effects, and bioavailability of bioactive chemicals can all be improved with these nanoscale delivery systems as compared to traditional drug administration techniques. According to Patra et al. (2018), nanocarriers are made to contain and safeguard bioactive molecules such as medicines, DNA, proteins, or nutrients inside their nano-sized structures. Lipids, polymers, or inorganic nanoparticles are only a few examples of the materials that can be used to create the nanocarriers. The ability to encapsulate and release the bioactive compounds in a controlled manner, stability, and biocompatibility are the criteria used to select these materials. The capability of nanocarriers to target particular bodily locations or cells is one of its main benefits. Nanocarriers can be designed to have specific targeting ligands or receptors that can recognize and bind to particular molecules or cells by surface functionalization or modification (Din et al., 2017). With the use of this targeting ability, bioactive chemicals can be precisely delivered to the desired site, such as tumor cells in cancer therapy or particular organs in targeted drug delivery. Bioactive substances can be stabilized and protected from deterioration by being enclosed inside nanocarriers. Drugs and enzymes are examples of bioactive chemicals that might degrade in the body or while being stored. Nanocarriers serve as protective barriers that delay premature deterioration and preserve the bioactivity of the substances they are encapsulating. The best therapeutic or functional effects of the bioactive compounds are ensured by this protection's capacity for controlled and prolonged release (Majumder et al., 2019).

Another significant advantage of nanocarriers is controlled release. The carriers' exact control over the release kinetics of the enclosed drugs is made possible by their nanoscale structures. Through a number of methods, such as diffusion, degradation, or stimuli-responsive behaviors, this regulated release can be accomplished. Nanocarriers offer extended therapeutic benefits, lower dose frequency, and lessen adverse effects related to high drug concentrations by adjusting the release rate and duration (Herdiana et al., 2022). Nanocarriers also have the potential to co-encapsulate a number of bioactive molecules, enabling combination treatments. With this strategy, the combination drugs can have synergistic effects and improve therapeutic results. For instance, medicines with various modes of action can be administered simultaneously to target several pathways in the treatment of cancer, increasing effectiveness and decreasing the risk of drug resistance (Eldar-Boock et al., 2013).

Nanocarriers have been investigated in the fields of agriculture and food science for the targeted delivery of bioactive substances, such as pesticides, herbicides, or fertilizers, to certain plant tissues or pests. This method of tailored administration lowers the amount of agrochemicals needed while minimizing environmental damage. However, it is crucial to take into account the safety and biocompatibility of nanocarriers when designing delivery systems. In the food industry, nanocarriers have been used to encapsulate bioactive compounds, such as vitamins, antioxidants, or probiotics, for improved nutrient delivery, enhanced stability, and controlled release in functional foods or dietary supplements. To ensure the safety of their usage in the delivery of bioactive agents, the potential toxicity or negative effects of the carrier materials should be thoroughly assessed (Guleria et al., 2022).

6.5 POTENTIAL RISKS ASSOCIATED WITH THE USE OF NANOMATERIALS IN FOOD BIOREMEDIATION

Numerous advantages, including enhanced food quality, safety, and functionality, are provided by the use of nanoparticles in culinary applications. To ensure the safe and responsible use of nanomaterials in the food sector, it is crucial to recognize and address any possible dangers related to their use. Certain nanoparticles may be hazardous and provide hazards to the environment and to human health. The toxicity of nanomaterials can be influenced by elements such as size, shape, surface charge, and chemical composition (Mohammad et al., 2022). According to several research, some nanoparticles may be able to cross cell membranes, collect in organs, or cause harmful biological reactions. To assure the safety of nanomaterials intended for culinary applications, it is essential to thoroughly evaluate their toxicity. The bioavailability and absorption of minerals and other food ingredients may be affected by nanomaterials (Mohammad et al., 2022). They can alter nutrition metabolism and absorption by interacting with food matrices, enzymes, and gastrointestinal tract elements. It is important to understand how nutritional absorption may be impacted by nanomaterials and how they can interfere with digestion and nutrient bioavailability (McClements and Xiao, 2017).

Food products using nanomaterials have the potential to cause allergic reactions in those who are prone to them. They might differ immunologically from traditional substances, which could trigger unanticipated allergic reactions. To safeguard the security of people with known allergies, rigorous allergenicity studies and labeling regulations are crucial (Gómez-Arribas et al., 2018). Ecological repercussions may result from the discharge of nanoparticles into the environment through wastewater, food waste, or disposal. The buildup of nanomaterials in soil, water bodies, and creatures, which can have an impact on ecosystems, is one aspect of the possible environmental impact. To reduce potential environmental dangers, adequate waste management procedures and environmental risk assessments are required. Consumers may have reservations about the usage of nanoparticles in food products (Martinez et al., 2020). Rejection may result from a lack of public understanding, transparency, and faith in the safety evaluation and risk management of nanomaterials. These worries can be addressed by open communication and consumer education on the advantages and safety issues of nanomaterials (Boholm and Larsson, 2019).

6.6 ADDRESSING CHALLENGES AND ENSURING SAFE IMPLEMENTATION OF NANOTECHNOLOGY IN THE FOOD INDUSTRY

The use of nanotechnology in the food sector has enormous potential to raise food quality, safety, and sustainability standards. To encourage customer confidence and trust, it is crucial to address the issues and guarantee the safe application of nanotechnology. For the purpose of identifying and reducing potential dangers related to the use of nanomaterials in food applications, thorough risk assessment and safety evaluation are essential. Amini et al. (2014) state that part of this process entails assessing the physicochemical characteristics, toxicity, bioavailability, and potential interactions of nanomaterials with food matrices and biological systems. It is important to create thorough safety evaluation processes that take into account the special characteristics and behavior of nanomaterials. For the use of nanotechnology in the food business to be safe, a clear and current regulatory framework is necessary. The evaluation, labeling, and safety of nanomaterials in food items need the establishment of norms and recommendations by regulatory organizations. To harmonize rules and enable the responsible commercialization of food products based on nanotechnology, cooperation bctween regulators, researchers, and industry stakeholders is required (Allan et al., 2021). For consumers to have a choice and make informed decisions, food goods containing nanomaterials must be clearly and transparently labeled. Nanomaterials should be properly disclosed on labels so that customers can make decisions based on their preferences or dietary requirements. To develop adoption of nanotechnology in the food sector and to foster trust among customers, regulators, and food makers, open communication must be maintained. For food products made with nanomaterials to be produced safely and consistently, GMP must be implemented specifically for this technology. The characterization of nanomaterials, production procedures, quality assurance, and risk management should all be covered under GMP rules. According to Siegrist and Keller (2011), following GMP guidelines reduces the chance of contamination, guarantees product uniformity, and aids in regulatory compliance. The potential environmental effects of nanoparticles employed in food production must be taken into account. This entails assessing their transfer, movement, and possible buildup in the environment. The identification of possible environmental hotspots and the implementation of sustainable practises can both be aided by life cycle analyses. Using effective waste management techniques is also essential to reducing environmental contamination. The key to increasing acceptance and knowledge of nanotechnology in the food business is encouraging consumer education and engagement. Any misunderstandings or worries can be addressed by educating customers about the advantages, safety concerns, and regulatory oversight of nanotechnology (Carroccio et al., 2022). A meaningful conversation can be sparked by open communication and the proactive pursuit of customer feedback. Addressing issues and guaranteeing the secure application of nanotechnology in the food industry requires cooperation among stakeholders, including researchers, business experts, regulators, and consumer advocacy organizations. Collective understanding is cultivated and evidence-based decision-making is made possible by the sharing of information, best practises, and research findings. To keep up with the most recent advances in nanotechnology and comprehend its possible effects on the food business, ongoing monitoring and research activities are essential (Boholm and Larsson, 2019). The establishment of revised safety guidelines and the improvement of risk assessment procedures are facilitated by monitoring developing nanomaterials, evaluating their safety, and analyzing their long-term impacts (Hristozov et al., 2016).

6.7 CONCLUSION

Food contamination issues may be addressed more effectively and precisely thanks to nanotechnology. Nanotechnology can minimize the discharge of pollutants into the environment, lessening their influence on ecosystems and natural resources. This is done through improving the removal of contaminants. Nanomaterial selective and focused remediation abilities help reduce the disruption of nontarget creatures and valuable elements in food systems, fostering ecological harmony and biodiversity. Nanotechnology improves monitoring and detection technologies for food safety, allowing for quick and real-time evaluation of food pollutants. The ability to identify and reduce possible threats to food safety is improved by this skill, enabling prompt intervention and preventive measures. Higher sensitivity and specificity provided by nanomaterial-based sensors and detection platforms enable the detection of pollutants at trace levels that could otherwise escape detection by conventional techniques. Nanotechnology aids in the creation of safer and better food items by ensuring early identification and action.

KEYWORDS

- **food preservation**
- **nanotechnology**
- **remediation**
- **shelf stability**
- **sustainability**

REFERENCES

Abatenh, E., Gizaw, B., Tsegaye, Z., & Wassie, M. (2017). The role of microorganisms in bioremediation—A review. *Open Journal of Environmental Biology, 2*(1), 038–046.

Ali, A. A., Altemimi, A. B., Alhelfi, N., & Ibrahim, S. A. (2020). Application of biosensors for detection of pathogenic food bacteria: A review. *Biosensors, 10*(6), 58.

Allan, J., Belz, S., Hoeveler, A., Hugas, M., Okuda, H., Patri, A., ... & Anklam, E. (2021). Regulatory landscape of nanotechnology and nanoplastics from a global perspective. *Regulatory Toxicology and Pharmacology, 122*, 104885.

Alori, E. T., Gabasawa, A. I., Elenwo, C. E., & Agbeyegbe, O. O. (2022). Bioremediation techniques as affected by limiting factors in soil environment. *Frontiers in Soil Science, 2*, 937186.

Alshawwa, S. Z., Kassem, A. A., Farid, R. M., Mostafa, S. K., & Labib, G. S. (2022). Nanocarrier drug delivery systems: Characterization, limitations, future perspectives and implementation of artificial intelligence. *Pharmaceutics, 14*(4), 883.

Amini, S. M., Gilaki, M., & Karchani, M. (2014). Safety of nanotechnology in food industries. *Electronic Physician, 6*(4), 962.

Arduini, F., Cinti, S., Scognamiglio, V., & Moscone, D. (2020). Nanomaterial-based sensors. In *Handbook of nanomaterials in analytical chemistry* (pp. 329–359). Elsevier.

Azubuike, C. C., Chikere, C. B., & Okpokwasili, G. C. (2016). Bioremediation techniques—Classification based on site of application: principles, advantages, limitations and prospects. *World Journal of Microbiology and Biotechnology, 32*, 1–18.

Bahrulolum, H., Nooraei, S., Javanshir, N., Tarrahimofrad, H., Mirbagheri, V. S., Easton, A. J., & Ahmadian, G. (2021). Green synthesis of metal nanoparticles using microorganisms and their application in the agrifood sector. *Journal of Nanobiotechnology, 19*(1), 1–26.

Bala, S., Garg, D., Thirumalesh, B. V., Sharma, M., Sridhar, K., Inbaraj, B. S., & Tripathi, M. (2022). Recent strategies for bioremediation of emerging pollutants: A review for a green and sustainable environment. *Toxics, 10*(8), 484.

Bayda, S., Adeel, M., Tuccinardi, T., Cordani, M., & Rizzolio, F. (2019). The history of nanoscience and nanotechnology: From chemical–physical applications to nanomedicine. *Molecules, 25*(1), 112.

Boholm, Å., & Larsson, S. (2019). What is the problem? A literature review on challenges facing the communication of nanotechnology to the public. *Journal of Nanoparticle Research, 21*, 1–21.

Carroccio, S. C., Scarfato, P., Bruno, E., Aprea, P., Dintcheva, N. T., & Filippone, G. (2022). Impact of nanoparticles on the environmental sustainability of polymer nanocomposites based on bioplastics or recycled plastics—A review of life-cycle assessment studies. *Journal of Cleaner Production, 335*, 130322.

Chen, T., & Dai, L. (2013). Carbon nanomaterials for high-performance supercapacitors. *Materials Today, 16*(7–8), 272–280.

Chopra, S., & Kumar, D. (2020). Ibuprofen as an emerging organic contaminant in environment, distribution and remediation. *Heliyon, 6*(6).

Curulli, A. (2020). Nanomaterials in electrochemical sensing area: Applications and challenges in food analysis. *Molecules, 25*(23), 5759.

Din, F. U., Aman, W., Ullah, I., Qureshi, O. S., Mustapha, O., Shafique, S., & Zeb, A. (2017). Effective use of nanocarriers as drug delivery systems for the treatment of selected tumors. *International Journal of Nanomedicine*, 7291–7309.

Edwards, C., & Lawton, L. A. (2009). Bioremediation of cyanotoxins. *Advances in Applied Microbiology, 67*, 109–129.

Eldar-Boock, A., Polyak, D., Scomparin, A., & Satchi-Fainaro, R. (2013). Nano-sized polymers and liposomes designed to deliver combination therapy for cancer. *Current Opinion in Biotechnology, 24*(4), 682–689.

Farooq, S. A., Raina, A., Mohan, S., Arvind Singh, R., Jayalakshmi, S., & Irfan Ul Haq, M. (2022). Nanostructured coatings: Review on processing techniques, corrosion behaviour and tribological performance. *Nanomaterials, 12*(8), 1323.

Fu, Z., & Xiang, J. (2020). Aptamer-functionalized nanoparticles in targeted delivery and cancer therapy. *International Journal of Molecular Sciences, 21*(23), 9123.

Gaviria-Arroyave, M. I., Cano, J. B., & Peñuela, G. A. (2020). Nanomaterial-based fluorescent biosensors for monitoring environmental pollutants: A critical review. *Talanta Open, 2*, 100006.

Gómez-Arribas, L. N., Benito-Peña, E., Hurtado-Sanchez, M. D. C., & Moreno-Bondi, M. C. (2018). Biosensing based on nanoparticles for food allergens detection. *Sensors, 18*(4), 1087.

Goutam Mukherjee, A., Ramesh Wanjari, U., Eladl, M. A., El-Sherbiny, M., Elsherbini, D. M. A., Sukumar, A., ... & Valsala Gopalakrishnan, A. (2022). Mixed contaminants: Occurrence, interactions, toxicity, detection, and remediation. *Molecules, 27*(8), 2577.

Guleria, G., Thakur, S., Shandilya, M., Sharma, S., Thakur, S., & Kalia, S. (2022). Nanotechnology for sustainable agro-food systems: The need and role of nanoparticles in protecting plants and improving crop productivity. *Plant Physiology and Biochemistry.*

Herdiana, Y., Wathoni, N., Shamsuddin, S., & Muchtaridi, M. (2022). Drug release study of the chitosan-based nanoparticles. *Heliyon.*

Hristozov, D., Gottardo, S., Semenzin, E., Oomen, A., Bos, P., Peijnenburg, W., ... & Marcomini, A. (2016). Frameworks and tools for risk assessment of manufactured nanomaterials. *Environment International, 95*, 36–53.

Javaid, M., Haleem, A., Singh, R. P., Rab, S., & Suman, R. (2021). Exploring the potential of nanosensors: A brief overview. *Sensors International, 2*, 100130.

Karigar, C. S., & Rao, S. S. (2011). Role of microbial enzymes in the bioremediation of pollutants: A review. *Enzyme Research, 2011.*

Khazaei, M., Hosseini, M. S., Haghighi, A. M., & Misaghi, M. (2023). Nanosensors and their applications in early diagnosis of cancer. *Sensing and Bio-Sensing Research,* 100569.

Kong, Y., Li, Y., Cui, X., Su, L., Ma, D., Lai, T., & Wang, Y. (2021). SnO2 nanostructured materials used as gas sensors for the detection of hazardous and flammable gases: A review. *Nano Materials Science.*

Kumar, H., Kuča, K., Bhatia, S. K., Saini, K., Kaushal, A., Verma, R., & Kumar, D. (2020). Applications of nanotechnology in sensor-based detection of foodborne pathogens. *Sensors, 20*(7), 1966.

Kumar, V., Shahi, S. K., & Singh, S. (2018). Bioremediation: An eco-sustainable approach for restoration of contaminated sites. *Microbial Bioprospecting for Sustainable Development,* 115–136.

Li, Z., He, Y., Sonne, C., Lam, S. S., Kirkham, M. B., Bolan, N., ... & Peng, W. (2022). A strategy for bioremediation of nuclear contaminants in the environment. *Environmental Pollution,* 120964.

Mahmoud, A. E. D., & Fawzy, M. (2021). Nanosensors and nanobiosensors for monitoring the environmental pollutants. *Waste Recycling Technologies for Nanomaterials Manufacturing,* 229–246.

Majumder, J., Taratula, O., & Minko, T. (2019). Nanocarrier-based systems for targeted and site specific therapeutic delivery. *Advanced Drug Delivery Reviews, 144*, 57–77.

Martínez, G., Merinero, M., Pérez-Aranda, M., Pérez-Soriano, E. M., Ortiz, T., Villamor, E., ... & Alcudia, A. (2020). Environmental impact of nanoparticles' application as an emerging technology: A review. *Materials, 14*(1), 166.

McClements, D. J., & Xiao, H. (2017). Is nano safe in foods? Establishing the factors impacting the gastrointestinal fate and toxicity of organic and inorganic food-grade nanoparticles. *NPJ Science of Food, 1*(1), 6.

Mohamed, E. F., & Awad, G. (2022). Development of nano-sensor and biosensor as an air pollution detection technique for the foreseeable future. In *Comprehensive analytical chemistry* (Vol. 99, pp. 163–188). Elsevier.

Mohammad, Z. H., Ahmad, F., Ibrahim, S. A., & Zaidi, S. (2022). Application of nanotechnology in different aspects of the food industry. *Discover Food, 2*(1), 12.

Murillo-Tovar, M. A., Saldarriaga-Noreña, H., & Saeid, A. (Eds.). (2021). Trace Metals in the Environment: New Approaches and Recent Advances.

Naresh, V., & Lee, N. (2021). A review on biosensors and recent development of nanostructured materials-enabled biosensors. *Sensors, 21*(4), 1109.

Ningthoujam, R., Singh, Y. D., Babu, P. J., Tirkey, A., Pradhan, S., & Sarma, M. (2022). Nanocatalyst in remediating environmental pollutants. *Chemical Physics Impact, 4*, 100064.

Patra, J. K., Das, G., Fraceto, L. F., Campos, E. V. R., Rodriguez-Torres, M. D. P., Acosta-Torres, L. S., ... & Shin, H. S. (2018). Nano based drug delivery systems: Recent developments and future prospects. *Journal of Nanobiotechnology, 16*(1), 1–33.

Pramanik, K., Maiti, T. K., & Mandal, N. C. (2021). Potential role of heavy metal-resistant plant growth-promoting rhizobacteria in the bioremediation of contaminated fields and enhancement of plant growth essential for sustainable agriculture. In *Recent advancement in microbial biotechnology* (pp. 357–385). Academic Press.

Seo, S. E., Tabei, F., Park, S. J., Askarian, B., Kim, K. H., Moallem, G., ... & Kwon, O. S. (2019). Smartphone with optical, physical, and electrochemical nanobiosensors. *Journal of Industrial and Engineering Chemistry, 77*, 1–11.

Shi, J., Votruba, A. R., Farokhzad, O. C., & Langer, R. (2010). Nanotechnology in drug delivery and tissue engineering: from discovery to applications. *Nano Letters, 10*(9), 3223–3230.

Siegrist, M., & Keller, C. (2011). Labeling of nanotechnology consumer products can influence risk and benefit perceptions. *Risk Analysis: An International Journal, 31*(11), 1762–1769.

Singh, Y., & Saxena, M. K. (2022). Insights into the recent advances in nano-bioremediation of pesticides from the contaminated soil. *Frontiers in Microbiology, 13*, 982611.

Song, J. H., Min, S. H., Kim, S. G., Cho, Y., & Ahn, S. H. (2022). Multi-functionalization strategies using nanomaterials: A review and case study in sensing applications. *International Journal of Precision Engineering and Manufacturing-Green Technology, 9*(1), 323–347.

Su, S., Wu, W., Gao, J., Lu, J., & Fan, C. (2012). Nanomaterials-based sensors for applications in environmental monitoring. *Journal of Materials Chemistry, 22*(35), 18101–18110.

Vaneev, A. N., Timoshenko, R. V., Gorelkin, P. V., Klyachko, N. L., Korchev, Y. E., & Erofeev, A. S. (2022). Nano-and microsensors for *in vivo* real-time electrochemical analysis: Present and future perspectives. *Nanomaterials, 12*(21), 3736.

Wang, Y., & Duncan, T. V. (2017). Nanoscale sensors for assuring the safety of food products. *Current Opinion in Biotechnology, 44*, 74–86.

Willner, M. R., & Vikesland, P. J. (2018). Nanomaterial enabled sensors for environmental contaminants. *Journal of Nanobiotechnology, 16*(1), 1–16.

CHAPTER 7

Challenges and Future Perspectives of Nano Technologically Derived Foods

RAMESH CHAND THAKUR[1], ABHINAY THAKUR[2], HARPREET KAUR[2*], CHITRA SHARMA[2], AKSHAY SHARMA[1], RENUKA SHARMA[1], and ASHISH KUMAR[3]

[1]*Department of Chemistry, Himachal Pradesh University, Shimla, Himachal Pradesh, India*

[2]*Department of Chemistry, School of Chemical Engineering and Physical Sciences, Lovely Professional University, Phagwara, Punjab, India*

[3]*NCE, Bihar Engineering University, Department of Science and Technology, Government of Bihar, India*

*Corresponding author

ABSTRACT

A vital agricultural regulating activity is monitored by nanotechnology, notably due to its microscopic size. Furthermore, a variety of conceivable advantages, like improved food quality and safety, a decrease in agricultural inputs, improved absorption of nanoscale nutrients from the soil, etc., enable the application of nanotechnology to be resonantly onerous and increase food productivity. With revolutionary nanotools for the management of rapid disease diagnosis, improving the ability of plants to absorb nutrients, and other applications, nanotechnology has the potential to advance the agriculture and food sector. Environmental resources, agriculture, and food all face difficulties related to susceptibility, sustainability, human health, and good quality of life. Therefore, more work needs to be done to raise public knowledge of the revolutionary food and agricultural

Nanotechnological Advancements in the Food Industry.
Ishrat Majid, Bababode Adesegun Kehinde, and Sheikh Shafat Hussain (Eds.)

DOI: 10.1201/9781774919750-7

products offered by nanotechnology. The prospect for nanotechnology uses in food is outlined in this chapter, which also begins with a consideration of the various problems that have resulted from these applications. It emphasizes that public perceptions and the eventual adoption or refusal of nanotech food items depend on what services they may provide, toward whom, and whether they may cause a hazard to consumers. It also addresses both the predicted benefits and concerns. In this context, it examines the possibility that some nanomaterials' unique chemical and physical properties could also result in previously undiscovered health risks. Additionally, the approaches for risk evaluation and the control of food-related nanotechnologies are discussed.

7.1 INTRODUCTION

In recent years, nanotechnology has emerged as a highly promising field with vast applications across various industries, including the food sector. Nanotechnology involves the manipulation and control of materials at the nanoscale level, typically ranging from 1 to 100 nm (Chau et al., 2007; Paramo et al., 2020; Shafiq et al., 2020). This level of precision enables scientists and engineers to harness unique properties and phenomena that occur at the nanoscale, opening up new avenues for innovation and advancement in various fields, including food production. Nanotechnologically derived foods, also known as nanofoods, are food products that incorporate nanoscale materials or utilize nanotechnology-based processes for their production (Jafarzadeh et al., 2020; Panpatte and Jhala, 2019). These technologies allow for precise control over food properties, such as texture, taste, and nutritional content. By manipulating food components at the nanoscale, scientists can enhance their functionality, improve their stability, and even create new flavors and textures. One of the key advantages of nanotechnologically derived foods lies in their potential to address pressing challenges faced by the global food system. With a rapidly growing population, diminishing natural resources, and increasing demands for safe, nutritious, and sustainable food, there is a critical need for innovative solutions. Nanotechnology offers a range of opportunities to revolutionize the way we produce, process, and consume food (Chausali et al., 2022; Siegrist et al., 2008). Enhancing food quality and safety is a primary focus of nanotechnologically derived foods. Nanoscale materials can be used to encapsulate bioactive compounds, such as vitamins, antioxidants, or probiotics, improving their stability and bioavailability. This targeted delivery system ensures that these valuable components are protected until they reach their intended destination in the body, maximizing their potential health benefits. Furthermore, nanosensors can be employed to detect and monitor pathogens, contaminants, and spoilage indicators in real time, enabling rapid response and mitigation strategies to ensure food safety (Soukoulis and Bohn, 2018). Nanotechnology offers a wide range of opportunities and benefits in the food industry, including improved food quality, safety, nutritional content, and extended shelf life (Fayaz et al., 2020). Here are some elaborations on how nanotechnology is being utilized in the food industry:

- **Nanoencapsulation:** Nanotechnology enables the encapsulation of bioactive compounds into nanoscale carriers. This process enhances the stability, solubility, and bioavailability of these compounds. For example, vitamins, antioxidants, and flavors can be encapsulated to protect them from degradation and improve their release properties. This technology allows for targeted delivery of these compounds to specific areas in the body, increasing their effectiveness (Mohammad et al., 2022; Siegrist et al., 2007). Nanoencapsulated vitamins, for instance, can be used to fortify food products, ensuring that consumers receive the desired nutritional benefits.

- **Food Packaging:** Nanotechnology has revolutionized food packaging by introducing advanced materials with enhanced barrier properties. Nanocomposites, which are films and coatings containing nanoparticles, provide excellent barrier properties against oxygen and moisture. This helps to extend the shelf life of packaged food by preventing spoilage and deterioration. Furthermore, incorporating antimicrobial nanoparticles into packaging materials can inhibit the growth of bacteria and fungi, reducing the risk of contamination and improving food safety.

- **Food Safety and Quality Monitoring:** Nanosensors and nanodevices are being developed for real-time monitoring of food quality and safety (Angelopoulou et al., 2022; Singh Sekhon, 2014; Sozer and Kokini, 2009). These nanoscale sensors can detect contaminants, pathogens, and spoilage indicators, providing rapid and sensitive detection methods. For instance, nanobiosensors can monitor the freshness of perishable foods by measuring specific biomarkers associated with spoilage. This technology enables early detection of potential issues, allowing for timely interventions and improved food safety.

- **Improved Texture and Sensory Attributes:** Nanotechnology can enhance the texture and sensory attributes of food products. By using nanoscale emulsions and nanostructured ingredients, the mouthfeel, stability, and appearance of food products can be modified. Nano-emulsion technology, for example, can improve the texture and stability of salad dressings, sauces, and beverages, resulting in a smoother and more appealing sensory experience for consumers.
- **Food Preservation:** Nanotechnology offers innovative approaches for food preservation. Antimicrobial nanoparticles can be incorporated into food packaging or directly applied to food surfaces, effectively inhibiting the growth of microorganisms and extending the shelf life of perishable products. Additionally, nano-based materials can be used for active food packaging, where they release antimicrobial agents or antioxidants to maintain food quality and freshness during storage.
- **Improved Nutrient Delivery:** Nanotechnology enables targeted and controlled nutrient delivery systems (Jagtiani, 2022; Lade and Gogle, 2019). Nanostructured carriers can protect sensitive nutrients during digestion, release them at specific sites in the body, and enhance their bioavailability. This approach can be used to improve the delivery of essential nutrients, such as vitamins and minerals, and facilitate their absorption by the body. This is particularly important in addressing nutrient deficiencies and improving overall health and well-being.
- **Food Processing and Production:** Nanotechnology can optimize food processing and production methods. Nanoscale materials can improve the efficiency of food processing equipment, such as filtration systems and membrane technologies, by enhancing separation and purification processes (Ogunkalu, 2019). Nanosensors and nanodevices can be utilized to monitor and control processing parameters, ensuring consistent quality and safety of the final food products. This technology also helps to reduce energy and

water consumption during food production, contributing to sustainability efforts. Figure 7.1 illustrates the utilization of nanotechnology in several fields of food industry (Primožiˇc, et al., 2021).

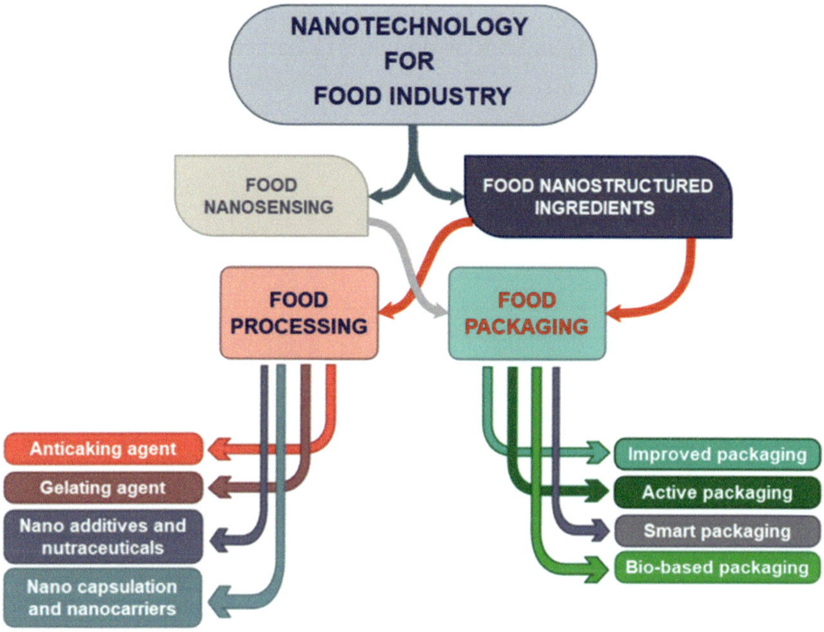

FIGURE 7.1 Application of nanotechnology in several domains of the food industry.

Source: Reprinted with permission from Primožiˇc, et al. (2021) Copyright © 2021 by the authors. Licensee MDPI, Basel, Switzerland. https://creativecommons.org/licenses/by/4.0/

While the utilization of nanotechnologically derived foods presents numerous advantages, it is important to ensure the responsible use of nanotechnology in food production. Additionally, sustainability is another crucial aspect that nanotechnologically derived foods can address. Nanotechnology offers opportunities to reduce food waste by extending the shelf life of perishable goods through improved packaging materials with enhanced barrier properties (Liskova et al., 2021; Valdés et al., 2009). Additionally, nanoscale additives can enhance the efficiency of agricultural practices, such as crop protection and nutrient delivery, leading to improved yields and reduced environmental impact. Despite the tremendous potential of nano technologically derived foods, several challenges need to be carefully considered and addressed. These challenges primarily revolve around regulatory frameworks, public perception and acceptance, scalability, and cost-effectiveness (Hosseini and Jafari, 2020). Ensuring the safety of nanoscale materials in food and establishing robust regulations for their use is crucial for consumer confidence. Figure 7.2 depicts the employment of nanotechnology in various domains (Khan et al., 2022).

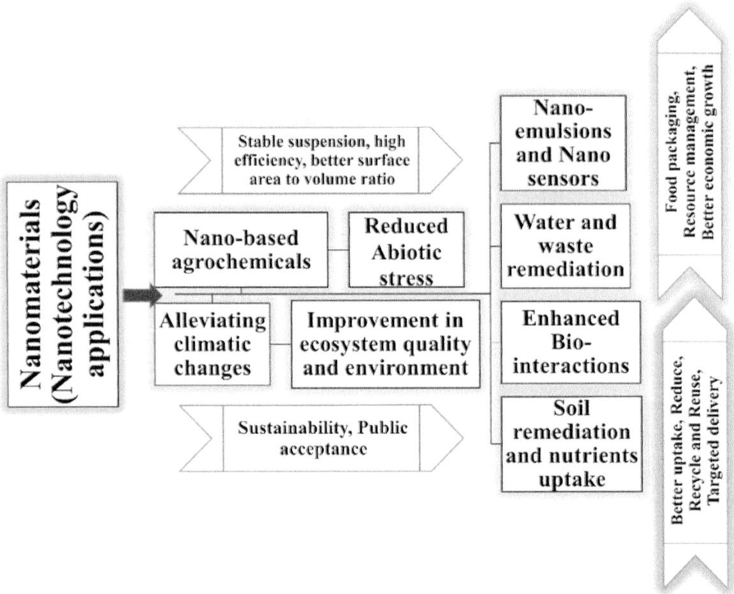

FIGURE 7.2 Application of nanomaterials in ecosystem and agricultural sustainability.

Source: Reprinted with permission from Khan et al. (2022) Copyright © 2022 by the authors. Licensee MDPI, Basel, Switzerland. https://creativecommons.org/licenses/by/4.0/

Public understanding and acceptance of nanotechnology in food are crucial for its successful integration into the global market. To achieve this, transparent communication and addressing concerns related to potential risks and ethical implications are essential. Several global examples highlight the efforts made to foster public understanding and acceptance of nanotechnology in the food industry.

- **European Union (EU):**

  The EU has been at the forefront of addressing public concerns and promoting transparent communication regarding nanotechnology in food. The European Food Safety Authority (EFSA) has developed guidelines for the risk assessment of nanomaterials in food, ensuring the safety of nanotechnologically derived food products (Cardoso-Ugarte et al., 2016). The EU-funded project NanoDiode engaged the public through interactive workshops, online platforms, and exhibitions to promote understanding and acceptance of nanotechnology. The project facilitated dialogue between stakeholders, fostering transparency and addressing concerns.

- **United States (US):**

  In the US, the National Nanotechnology Initiative (NNI) has played a significant role in addressing public concerns related to nanotechnology, including its applications in the food industry. The NNI supports research and public engagement activities to enhance understanding and acceptance (Hou et al., 2022; Rashidi and Khosravi-Darani, 2011). The Food and Drug Administration (FDA) regulates nanotechnology in food, providing guidance to manufacturers and conducting risk assessments to ensure the safety of nanotechnologically derived food products.

- **Australia:**

  Australia has established the Australian Centre for Nanoscale BioPhotonics (CNBP), a research consortium focusing on the development and application of nanotechnologies in various fields, including food and agriculture (Baghi et al., 2022; Jain et al., 2018). The CNBP engages in public outreach programs, such as science festivals and public lectures, to promote understanding and acceptance of nanotechnology. The Australian government also emphasizes the importance of transparent communication and encourages public dialogue on emerging technologies, including nanotechnology.

- **Canada:**

  Canada has taken a proactive approach to address public concerns and foster public acceptance of nanotechnology. The Canadian Food Inspection Agency (CFIA) has implemented a regulatory framework to ensure the safety of nanotechnology applications in food. The CFIA collaborates with other regulatory bodies and engages in public consultations to gather input and address concerns. The government has also invested in research and development programs to support the responsible development and use of nanotechnologies in food production.

- **Japan:**

  Japan has been at the forefront of nanotechnology research and its application in various industries, including food. The Japanese government promotes public understanding and acceptance through initiatives such as the NanoSafety Forum for Citizens, which aims to engage the public in discussions on the safety and ethical aspects of nanotechnology. The Ministry of Health, Labour, and Welfare has established guidelines for the safety assessment of nanomaterials used in food and promotes research collaborations to enhance knowledge and safety measures (Prasad et al., 2022). These global examples demonstrate the efforts made by regulatory agencies, industry stakeholders, and governments to foster public understanding and acceptance of nanotechnology in food. Transparent communication, engagement with the public, and addressing concerns through research, regulations, and risk assessments are essential for building trust and ensuring responsible development and use of nanotechnologically derived food products.

The objective of this book chapter is to comprehensively examine the challenges and future perspectives associated with nano technologically derived foods. The chapter aims to provide insights into the responsible integration and development of nanotechnology in the food industry. The specific objectives include providing an overview of nanotechnology and its applications in the food sector, exploring the regulatory challenges of nano technologically derived foods, investigating the public perception and acceptance of these foods, assessing the scalability and cost-effectiveness of nano technologically derived food production, exploring the future perspectives and potential applications of these foods, and discussing the ethical and societal implications of nano technologically derived foods.

The chapter begins by discussing the fundamental concepts of nanotechnology and highlighting its potential benefits in terms of improved food quality, safety, and functionality. It then delves into the regulatory challenges associated with nano technologically derived foods, analyzing current regulatory frameworks and emphasizing the importance of robust regulations for consumer safety and confidence. The chapter also explores public attitudes, concerns, and knowledge regarding nanotechnology in food, and proposes strategies to address these concerns and promote consumer acceptance. Additionally, the chapter examines the challenges and opportunities of scaling up nanotechnology processes for large-scale food production, considering factors such as safety, quality, and economic viability. It explores emerging trends and innovations in nanotechnology for food production, including nanoencapsulation for targeted nutrient delivery, nanosensors for real-time monitoring of food quality, and novel nanomaterials for improved packaging and extended shelf life.

Furthermore, the chapter addresses the ethical considerations and potential societal impacts associated with nano technologically derived foods, focusing on issues of equity, accessibility, and environmental sustainability. By addressing these objectives, the chapter aims to provide a comprehensive understanding of the challenges and future perspectives of nano technologically derived foods. It aims to inform decision-making, policy development, and future research in the field, ultimately facilitating the responsible integration of nanotechnology into the food industry to enhance food quality, safety, and sustainability.

7.2 REGULATORY CHALLENGES

7.2.1 NANOPARTICLES IN FOOD: POTENTIAL RISKS AND SAFETY CONCERNS

The utilization of nanotechnology in food production has gained significant attention due to its potential to revolutionize the food industry. However, the introduction of nanoscale materials and processes into food products also brings forth unique challenges and safety concerns that must be carefully addressed through robust regulatory frameworks. One of the primary concerns is the presence of nanoparticles in food and their potential risks to human health. Nanoparticles possess distinct physicochemical properties compared to bulk materials, primarily due to their small size and high surface area-to-volume ratio (Calabrese and Ferranti, 2018; Gutiérrez et al., 2017; Vijayakumar et al., 2022; Weiss et al., 2006). These properties can significantly influence their behavior and interactions within biological systems upon ingestion. Scientific research has indicated that certain nanoparticles may exhibit toxic effects. These effects can include inflammation, DNA damage, or the ability to cross biological barriers such as the blood–brain barrier or placental barrier. The reactivity of nanoparticles, arising from their small size, can result in enhanced interactions with biological molecules and cellular components (Babu, 2022; Sahoo et al., 2022). This increased reactivity may lead to adverse biological responses, raising concerns about their safety when present in food products. Figure 7.3 illustrates the type of nanocarriers used in food industry (Khezerlou et al., 2021).

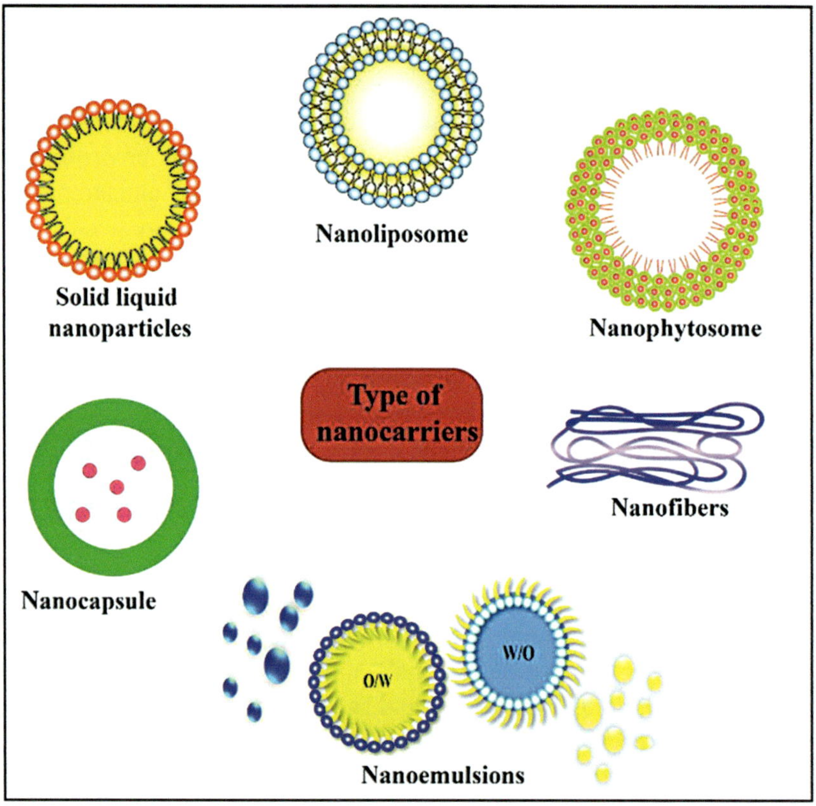

FIGURE 7.3 Types of nanocarriers for use in food packaging material.

Source: Reprinted with permission from Khezerlou et al. (2021) Copyright © 2021 by the authors. Licensee MDPI, Basel, Switzerland. https://creativecommons.org/licenses/by/4.0/

To ensure the safety of nanoparticles in food, thorough risk assessments are imperative. These assessments involve evaluating the potential toxicity of nanoparticles, studying their behavior and fate in various biological systems, and establishing safe exposure limits. Robust toxicological studies are conducted to understand the mechanisms of nanoparticle interactions with biological systems, identifying any potential adverse effects and determining safe levels of exposure for consumers (Prasad et al., 2017). Furthermore, comprehensive studies are necessary to assess the long-term effects of nanoparticle ingestion. Nanoparticles may exhibit different biodistribution and bioaccumulation patterns compared to larger particles. Understanding how nanoparticles are metabolized and eliminated from the body is critical to evaluating their safety and potential for long-term health implications (Khezerlou et al., 2021). The establishment of robust regulatory frameworks is essential to address these safety concerns effectively. Regulatory agencies need to develop clear guidelines and standards for the characterization, safety assessment, and labeling of nano-based food products. These guidelines should encompass methodologies for nanoparticle detection, quantification, and toxicological testing. Standardized protocols ensure consistency and reliability in risk assessments, facilitating accurate evaluations of nanoparticle safety in food.

Moreover, effective risk communication is crucial to address public concerns and build consumer confidence. Transparent and science-based communication regarding the safety and benefits of nano technologically derived foods is essential for fostering trust. Regulatory agencies should engage in proactive communication strategies, providing accurate information and addressing any misconceptions or fears associated with nanoparticles in food.

7.2.2 NEED FOR ROBUST REGULATORY FRAMEWORKS

In recent years, the need for robust regulatory frameworks to address the challenges associated with nanotechnology in food production has become increasingly evident. Data findings at the global level highlight the necessity of dedicated regulations and guidelines to ensure the safe integration of nano technologically derived foods into the market. A study conducted by the European Commission in 2018 assessed the current regulatory landscape for nanomaterials in food and feed in several countries. The findings revealed a lack of harmonization and consistency in the regulatory approaches, with variations in definitions, measurement techniques, and risk assessment methodologies. This lack of uniformity hinders the effective regulation of nano-based food products and poses challenges for industry compliance. Furthermore, data from a survey conducted by the Food Standards Agency in the United Kingdom highlighted consumer concerns regarding the safety and transparency of nanotechnology in food. The survey found that 64% of respondents were worried about the use of nanotechnology in food, primarily due to a lack of information and understanding. This underscores the importance of clear labeling requirements for nano-based food products to provide consumers with the necessary information to make informed choices.

The importance of collaboration between regulatory agencies, scientific experts, and industry stakeholders is supported by various initiatives and research collaborations worldwide. For example, the Global Harmonization Initiative (GHI) has been working toward harmonizing regulations and standards for nanotechnology applications, including food. The GHI promotes collaboration and knowledge-sharing among regulators, scientists, and industry professionals to address the challenges associated with nanotechnology and ensure consumer safety. Moreover, organizations such as the International Organization for Standardization (ISO) have developed specific standards for nanotechnologies, including those applicable to the food industry (Fadiji et al., 2022; Khan et al., 2022). These standards provide guidelines for the characterization, measurement, and labeling of nanomaterials, contributing to the development of robust regulatory frameworks. The need for standardized methods and protocols is also evident in the research community. Scientists and researchers have been working on developing standardized toxicity testing protocols for nanomaterials to ensure reliable and consistent risk assessments. These efforts aim to provide regulatory agencies with scientifically sound data for evaluating the safety of nano-based food products.

7.2.3 ENSURING CONSUMER SAFETY AND CONFIDENCE

Ensuring consumer safety and confidence is paramount in the regulation of nano technologically derived foods. The introduction of novel technologies can sometimes create uncertainty and skepticism among consumers (Lamri et al., 2021; Primoži˘c, et al., 2021). Therefore, it is crucial to establish transparent and science-based approaches to address safety concerns and build consumer trust. Effective risk communication plays a pivotal role in ensuring consumer confidence. Regulatory agencies should engage in proactive and transparent communication to provide accurate information about the safety and benefits of nano technologically derived foods. Open dialogue can help dispel misconceptions, address concerns, and foster a better understanding of the risks and benefits associated with these products (Kaur et al., 2020, 2021, 2022; Nidhi, 2022; Thakur et al., 2020). Consumer engagement is another important aspect of ensuring safety and confidence. Including consumer representatives in decision-making processes and inviting public input on regulatory matters can enhance transparency and democratic participation. Consumer input can help shape regulations, ensuring that they align with public expectations and values. Figure 7.4 depicts the essential steps that could be processed for ensuring consumer safety and confidence (Shafiq et al., 2020).

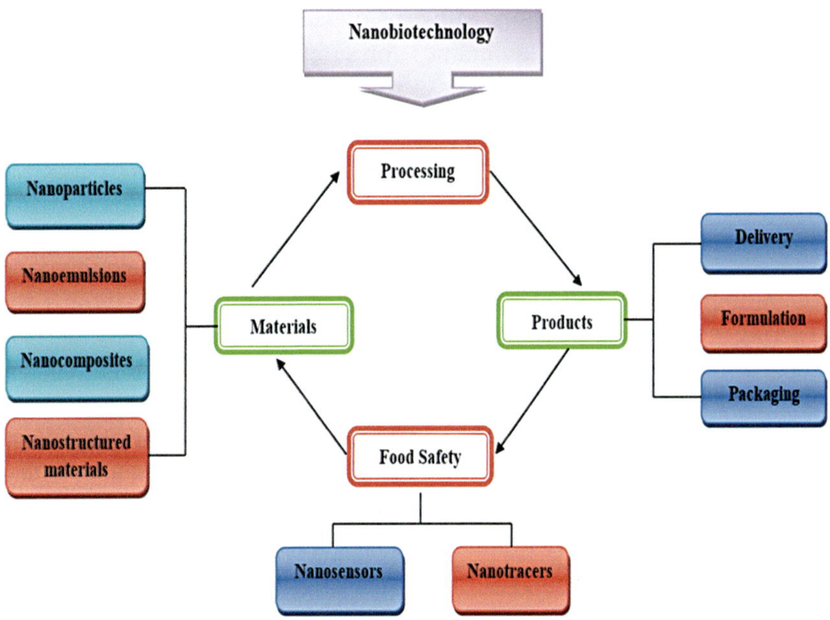

FIGURE 7.4 Role of nanotechnology in various sectors of the food industry.

Source: Reprinted with permission from Shafiq et al. (2020) Copyright © 2021 by the authors. Licensee MDPI, Basel, Switzerland. https://creativecommons.org/licenses/by/4.0/

Surveillance and monitoring programs are vital for assessing the safety of nano-based food products in the market. Regulatory agencies around the world have taken various measures to address safety concerns and build consumer trust. One example is the European Union (EU), which has implemented regulations requiring the labeling of food products containing engineered nanomaterials, ensuring transparency and informed consumer choice (Bashir et al., 2019; Ganjoo et al., 2022; Thakur and Kumar, 2021, 2022; Thakur et al., 2022a, 2022b, 2022c, 2022d, 2022e, 2022f, 2023). In the United States, the Food and Drug Administration (FDA) has established a voluntary regulatory program for the safety assessment of nanotechnology products, encouraging manufacturers to consult with the agency before market entry. Similarly, Australia's Food Standards Australia New Zealand (FSANZ) has developed a specific standard for novel foods, including nanotechnology-derived foods, ensuring comprehensive safety assessments before approval. Canada's Health Canada conducts safety assessments and provides guidance to manufacturers, while Japan's Ministry of Health, Labour, and Welfare (MHLW) has established guidelines for safety assessment and labeling of nano-based food products. These regulatory efforts are supported by international cooperation and harmonization initiatives. The International Organization for Standardization (ISO) has developed globally recognized standards for nanomaterial characterization and measurement. These examples demonstrate proactive risk communication, consumer engagement, surveillance and monitoring programs, and international collaboration, all aimed at ensuring consumer safety, promoting transparency, and fostering confidence in the regulation of nano technologically derived foods. International cooperation and harmonization of regulations are essential in a globalized food market. Collaboration between regulatory agencies across different countries facilitates the exchange of information, best practices, and harmonization of safety standards. This enables a consistent and coherent regulatory approach, avoids trade barriers, and enhances consumer protection on a global scale.

7.3 PUBLIC PERCEPTION AND ACCEPTANCE

Public perception and acceptance play a critical role in the successful integration of nanotechnology in the food industry. While nanotechnologically derived foods offer numerous potential benefits, public concerns and perceptions can influence consumer acceptance and adoption of these products. Understanding the factors influencing public acceptance and addressing their concerns are crucial for building trust and facilitating the responsible development and implementation of nano-based food products.

7.3.1 NANOTECHNOLOGY IN FOOD: PUBLIC CONCERNS AND PERCEPTION

Public concerns regarding nanotechnology in food can stem from various factors, including limited awareness and knowledge, uncertainty about the potential risks and benefits, and ethical and moral considerations. A lack of understanding of nanotechnology and its applications can lead to skepticism and resistance. Misinformation or exaggerated claims in the media can also contribute to public apprehension. Public perception regarding the safety of nano-based food products is a significant concern. Some individuals may be wary of consuming food that contains nanomaterials due to fears of potential health risks. There is a need for transparent and credible information regarding the safety assessment and risk management of these products to address these concerns.

Numerous countries and organizations have made efforts to address public concerns and enhance acceptance of nanotechnologically derived foods. For example, the European Union (EU) has implemented regulations to ensure the safety of nanomaterials in food. The EU's Novel Food Regulation requires companies to seek approval before placing nano-based food products on the market. The regulation also mandates labeling requirements to inform consumers about the presence of nanomaterials in food. In the United States, the Food and Drug Administration (FDA) has issued guidance on the safety assessment of nanomaterials in food. The FDA encourages manufacturers to consult with the agency when developing nano-based food products and provides recommendations for conducting safety evaluations. The agency also collaborates with other organizations to enhance knowledge and understanding of nanotechnology's potential risks and benefits.

Australia's regulatory agency, Food Standards Australia New Zealand (FSANZ), has established guidelines for the safety assessment and regulation of nanotechnologically derived foods. FSANZ conducts thorough risk assessments and requires mandatory labeling of foods containing novel nanomaterials. Singapore has taken a proactive approach to address public concerns (Canteli et al., 2014). The country's Agency for Science, Technology, and Research (A*STAR) has collaborated with various stakeholders to establish the Nano and Materials Programme. This program aims to promote the responsible development and use of nanomaterials, including in the food industry. It focuses on research, risk assessment, and public engagement to enhance understanding and acceptance of nanotechnology. In Japan, the Ministry of Health, Labour, and Welfare (MHLW) has implemented guidelines for the safety assessment of nanomaterials in food. The MHLW conducts comprehensive evaluations, including toxicity testing, to ensure the safety of nano-based food products before they are approved for commercialization.

Efforts to address public concerns and enhance acceptance of nanotechnologically derived foods extend beyond regulatory measures (Granato et al., 2014; Lopes et al., 2021). For instance, the NanoRiskCat project in Switzerland aims to provide a platform for the systematic assessment of nanomaterials in various applications, including food. The project involves collaboration between academic institutions, industry, and government agencies to evaluate the potential risks associated with nanomaterials and develop risk assessment tools. These examples demonstrate the global recognition of the need to address public concerns and ensure the responsible development and use of nanotechnologically derived foods. Through regulatory measures, risk assessment protocols, transparent communication, and public engagement initiatives, efforts are being made to enhance understanding, build trust, and promote consumer acceptance of these innovative food products (Bashir et al., 2020; Kumar and Thakur, 2020; Parveen et al., 2020; Sandilya et al., 2021; Sharma et al., 2014; Thakur and Kumar, 2023; Thakur et al., 2021; Verma et al., 2023).

7.3.2 FACTORS INFLUENCING PUBLIC ACCEPTANCE

Several factors influence public acceptance of nanotechnologically derived foods. Trust in regulatory bodies and the transparency of the regulatory process are crucial determinants. Consumers look to regulatory agencies for assurance that these products are safe for consumption and meet appropriate standards. Therefore, effective communication and engagement between regulatory authorities, industry, and the public are essential.

Perceived benefits of nano technologically derived foods, such as improved nutritional content, enhanced sensory attributes, and extended shelf life, can positively influence public acceptance. Consumer perception of the added value and potential advantages over conventional food products can play a significant role in shaping their acceptance (Asimakopoulos et al., 2021; Luo et al., 2020). Socio-cultural factors and individual values also impact public acceptance. Attitudes toward technology, perceptions of naturalness, and concerns about the ethical implications of nanotechnology can influence consumer acceptance. Understanding these socio-cultural factors is important for tailoring communication strategies and addressing specific concerns within different populations.

The European Food Safety Authority (EFSA) conducted a study on consumer perceptions of nanotechnology in food. The study found that consumers' knowledge and understanding of nanotechnology were limited, and they expressed concerns about the potential risks. As a response, EFSA has developed a framework for risk communication and engagement with consumers, aiming to address their concerns and provide accurate information about nanotechnologically derived foods. The Nanotechnology Consumer Products Inventory (CPI) in the United States provides a database of consumer products that use nanotechnology, including food and beverages. The inventory offers information on the types of nanomaterials used, potential exposure routes, and regulatory status. This transparency helps consumers make informed decisions and fosters trust in the industry (Oliveira et al., 2015; Pauley and Maskell, 2017).

In Australia, the Commonwealth Scientific and Industrial Research Organisation (CSIRO) conducted a study to assess consumer perceptions of nanotechnology in food and packaging (Shibata et al., 2009; Simate et al., 2012; Xing et al., 2019). The study found that while consumers were generally unfamiliar with nanotechnology, they expressed concerns about its potential risks. As a result, CSIRO initiated public engagement activities, including workshops and information sessions, to enhance consumer understanding and address their concerns. The Danish Consumer Council developed guidelines for responsible communication about nanotechnologically derived foods. These guidelines emphasize transparency, accurate information, and clear labeling. They encourage food producers and retailers to provide understandable and accessible information about the use of nanotechnology in food products to build consumer trust (Onuki et al., 2015). In Japan, the National Institute for Materials Science

(NIMS) conducted surveys to investigate public perceptions and attitudes toward nanotechnology. The surveys found that consumers were generally cautious about nanotechnology in food and expressed concerns about potential health risks. In response, NIMS has been actively engaging with the public through workshops and exhibitions to promote understanding and dialogue. These examples highlight the importance of addressing public concerns, promoting transparency, and tailoring communication strategies to enhance consumer acceptance of nanotechnologically derived foods. By engaging with consumers, providing accurate information, and considering socio-cultural factors, stakeholders can work toward building trust and fostering acceptance of these innovative food products.

7.3.3 ADDRESSING PUBLIC CONCERNS AND BUILDING TRUST

Public perception and acceptance play a significant role in the successful integration of nanotechnology in the food industry. As nanotechnologically derived foods continue to emerge, it is essential to address public concerns, provide accurate information, and engage in transparent communication to build trust among consumers. This article will elaborate on various strategies and initiatives that have been implemented globally to address public concerns and foster acceptance of nanotechnologically derived foods. By promoting open dialogue, improving public awareness, prioritizing risk communication, and fostering collaboration, stakeholders can work toward building trust and facilitating the responsible development and acceptance of these innovative food products. One of the fundamental approaches to address public concerns and build trust is to promote open and transparent dialogue between stakeholders and the public. Engaging with consumers through public consultations, focus groups, and online platforms enables direct interaction and allows for the exchange of information and perspectives. This participatory approach not only helps address concerns and clarify misconceptions but also empowers consumers by giving them a voice in decision-making processes. Involving the public in discussions related to the regulation and safety assessment of nanotechnologically derived foods ensures that their perspectives and values are taken into account, thus contributing to a more inclusive and transparent regulatory framework.

Improving public understanding of nanotechnology and its applications in food is crucial for addressing concerns and building trust.

Education and awareness campaigns can play a significant role in providing clear, accessible, and scientifically accurate information to the public. These campaigns should focus on the benefits of nanotechnologically derived foods, the safety assessment procedures, and the regulatory oversight in place. By disseminating accurate information, dispelling misconceptions, and fostering informed decision-making, stakeholders can help enhance public acceptance of these innovative food products. Regulatory agencies have a vital role in addressing public concerns and ensuring the safety of nanotechnologically derived foods. Prioritizing risk communication and providing transparent information about the safety assessment and risk management of these products is essential. This includes establishing clear labeling requirements to ensure consumer awareness and choice (Li et al., 2021; Wu et al., 2021). Labels should not only indicate the presence of nanomaterials but also provide information regarding their nature, function, and potential risks. By ensuring transparency in labeling, consumers can make informed decisions based on their preferences and concerns. Additionally, regulatory bodies should establish mechanisms for ongoing monitoring and post-market surveillance to promptly address any safety concerns that may arise, further enhancing consumer confidence in these products.

Building trust requires collaboration among regulatory agencies, industry stakeholders, scientific experts, and consumer representatives. Public–private partnerships can facilitate knowledge sharing, research collaboration, and the development of best practices. By working together, stakeholders can address emerging issues, share expertise, and foster a collective responsibility toward the responsible development and acceptance of nanotechnologically derived foods. International cooperation and harmonization of regulations also play a crucial role in building global trust (Leonardo et al., 2017). Collaboration between regulatory agencies across different countries enables the exchange of information, best practices, and the harmonization of safety standards. This promotes consistency and coherence in the regulatory approach, avoids trade barriers, and enhances consumer protection on a global scale. Several countries have implemented initiatives to address public concerns and enhance acceptance of nanotechnologically derived foods. These examples illustrate the diverse strategies employed to foster understanding and build trust:

- **European Union (EU):** The NanoDiode project was launched to engage the public in dialogue and deliberation on nanotechnology-related issues. The project involved interactive workshops, online platforms, and exhibitions to promote public understanding and acceptance of nanotechnology.

- **United States:** The National Nanotechnology Initiative (NNI) established the Nanotechnology Environmental and Health Implications (NEHI) Working Group to assess the environmental, health, and safety implications of nanotechnology. The NNI supports research and public engagement activities to address societal concerns and promote responsible development.
- **Australia:** The Commonwealth Scientific and Industrial Research Organisation (CSIRO) conducted a study to assess consumer perceptions of nanotechnology in food and packaging. Based on the findings, CSIRO initiated public engagement activities, including workshops and information sessions, to enhance consumer understanding and address their concerns.
- **Denmark:** The Danish Consumer Council developed guidelines for responsible communication about nanotechnologically derived foods. These guidelines emphasize transparency, accurate information, and clear labeling. They encourage food producers and retailers to provide understandable and accessible information about the use of nanotechnology in food products to build consumer trust.
- **Japan:** The National Institute for Materials Science (NIMS) conducted surveys to investigate public perceptions and attitudes toward nanotechnology. In response to public concerns, NIMS has been actively engaging with the public through workshops and exhibitions to promote understanding and dialogue.

These examples demonstrate the importance of proactive communication, public engagement, education, and collaborative efforts among regulatory agencies, industry stakeholders, and the public. By fostering a better understanding of nanotechnology and its benefits, addressing safety concerns, and involving the public in decision-making processes, stakeholders can work toward building trust and fostering the responsible development and acceptance of nanotechnologically derived foods.

7.4 SCALABILITY AND COST-EFFECTIVENESS

Scaling up nanotechnology processes for large-scale food production presents several challenges that need to be addressed to ensure efficient and cost-effective manufacturing. One such challenge is the development of scalable production methods that can produce nanomaterials in large quantities while maintaining consistent quality. An example of successful scaling up is the production of nanoemulsions, which are widely used in the food industry for encapsulating bioactive compounds and improving product stability. The company Blue California developed a large-scale production process for nanoemulsions using high-pressure homogenization. By optimizing the parameters and equipment, they were able to achieve efficient and cost-effective production, allowing for the widespread incorporation of nanoemulsions in various food and beverage products.

Another aspect to consider is the sourcing of raw materials. Scaling up may require access to larger quantities of specific raw materials or the development of cost-effective alternatives. For instance, in the production of nanostructured lipid carriers (NLCs), which are used for encapsulating lipophilic bioactive compounds, researchers have explored the use of locally available and sustainable raw materials. In Malaysia, researchers utilized palm oil, a widely available and cost-effective raw material, to produce NLCs on a large scale. This approach not only ensured scalability but also contributed to the economic viability of the process. Collaboration between different stakeholders is crucial for addressing scalability challenges. The European Union-funded project NanoPack aimed to develop and demonstrate the industrial viability of nanotechnology-enabled packaging solutions. The project involved collaboration between academic institutions, research organizations, and industry partners to develop scalable processes for the production of antimicrobial nanopackaging materials. By working together, they successfully scaled up the production of nanocomposite films with antimicrobial properties, paving the way for their integration into large-scale food packaging applications.

7.4.1 SCALING UP NANOTECHNOLOGY PROCESSES FOR LARGE-SCALE FOOD PRODUCTION

Scaling up nanotechnology processes for large-scale food production is a multifaceted challenge that involves several key considerations (Alzeer and Abou Hadeed, 2016; Golovacheva et al., 2021). Process optimization is crucial in developing scalable production methods that efficiently produce nanomaterials while maintaining consistent quality. This may involve modifying existing manufacturing processes or developing new approaches tailored for large-scale production. For example, continuous manufacturing techniques like microfluidics can enable the continuous production of nanoparticles at higher volumes with improved control (Marynchenko et al., 2017; Raspor and Goranovič, 2008). Ensuring a stable supply chain and securing reliable sources of raw materials is vital for achieving scalability. Collaborating with researchers, industry stakeholders, and technology providers facilitates knowledge sharing and resource pooling, accelerating the development of scalable nanotechnology processes. Examples include Malaysia's use of locally available and sustainable raw materials like palm oil for large-scale production. Establishing an efficient supply chain involves coordinating the sourcing, transportation, and distribution of raw materials, as well as ensuring the availability of equipment and resources. Moreover, maintaining process safety and quality assurance is essential when scaling up nanotechnology processes. Implementing rigorous testing protocols, quality control measures, and adhering to regulatory guidelines ensure the safety and quality of nanomaterials and final food products. Collaboration and technology transfer between academia and industry support the application of research findings in commercial settings. Initiatives like the NanoPack project in Europe and the National Nanotechnology Initiative in the United States promote collaboration, knowledge exchange, and the advancement of nanotechnology applications. By addressing these challenges and leveraging global examples and collaborative efforts, the food industry can harness the full potential of nanotechnology in delivering safe and cost-effective products to meet consumer demands. Several global examples highlight the efforts and advancements in scaling up nanotechnology processes for large-scale food production:

- **NanoPack Project (Europe):** The NanoPack project, funded by the European Union's Horizon 2020 program, aimed to develop and demonstrate the industrial viability of nanotechnology-enabled packaging solutions. The project involved collaboration between academic institutions, research organizations, and industry partners. They successfully scaled up the production of antimicrobial nanocomposite films, which extended the shelf life of perishable food products. By utilizing scalable manufacturing processes, NanoPack contributed to the integration of nanotechnology into large-scale food packaging applications.

- **Blue California (United States):** Blue California is a company that developed a large-scale production process for nanoemulsions, which are used in various food and beverage applications. They optimized high-pressure homogenization parameters and equipment to achieve efficient and cost-effective production. Through their scalable manufacturing process, Blue California has enabled the widespread incorporation of nanoemulsions into products such as beverages, sauces, and dressings.
- **Palm Oil-based Nanostructured Lipid Carriers (Malaysia):** Researchers in Malaysia explored the use of locally available and cost-effective raw materials for the large-scale production of nanostructured lipid carriers (NLCs). By utilizing palm oil, a widely available and sustainable raw material, they successfully scaled up the production of NLCs. This approach not only addressed scalability challenges but also contributed to the economic viability of the process, making it more accessible for the food industry.
- **Nanocoating Technologies (Israel):** Nanocoating Technologies, an Israeli company, developed a scalable manufacturing process for nanocoatings used in food packaging. Their nanocoatings provide a barrier against oxygen, moisture, and microbial contamination, extending the shelf life of food products. By optimizing the manufacturing process and collaborating with packaging manufacturers, Nanocoating Technologies achieved large-scale production of nanocoated packaging materials, demonstrating the commercial viability of their technology.

7.4.2 MAINTAINING SAFETY AND QUALITY AT SCALE

As nanotechnology processes are scaled up for large-scale food production, ensuring safety and maintaining quality become increasingly critical. The unique properties and characteristics of nanomaterials require thorough evaluation and assessment to ensure that their use in food production does not compromise consumer safety. Regulatory compliance, characterization, risk assessment, quality control, and personnel training are all essential aspects to consider in maintaining safety and quality in the production of nanotechnologically derived foods (Onuki et al., 2016; Shibata et al., 2009). Characterization and risk assessment of nanomaterials play a pivotal role in maintaining safety. As production scales up, it becomes necessary to establish reliable and standardized methods for characterizing and assessing the properties, behavior, and potential risks of nanomaterials. This involves evaluating the physicochemical properties, toxicological profiles, and potential interactions with food matrices. Understanding the behavior and potential risks of nanomaterials is crucial for ensuring that the final food products are safe for consumption. Quality control and quality assurance processes must be implemented throughout the production chain to monitor and verify the consistency and integrity of the final products. This includes implementing rigorous testing protocols, product traceability systems, and adherence to regulatory standards. Robust quality management systems, such as Good Manufacturing Practices (GMP), are crucial for ensuring that safety and quality are maintained at every stage of the production process. Regular audits and inspections can help identify any deviations from safety and quality requirements and allow for timely corrective actions (Cai et al., 2016; Siríštová et al., 2012).

Training and education programs for personnel involved in large-scale production are essential to ensure that they have the necessary knowledge and skills to handle nanomaterials safely and maintain the desired quality standards. This includes training on proper handling, storage, and disposal of nanomaterials, as well as understanding the potential risks associated with their use. Ongoing training and education programs can help keep personnel up-to-date with the latest safety practices and regulations. Collaboration between regulatory agencies, industry, and scientific communities is crucial for developing comprehensive safety guidelines and regulatory frameworks for nanotechnologically derived foods. Regulatory bodies play a vital role in establishing safety standards and ensuring compliance with regulations. Industry stakeholders and scientific communities can contribute their expertise and knowledge to inform regulatory decision-making processes. By working together, these entities can develop effective guidelines that encompass the entire production chain and address the unique challenges associated with scaling up nanotechnology processes. Global examples demonstrate efforts to ensure safety and quality in the production of nanotechnologically derived foods.

For instance, the European Union has established a regulatory framework for novel foods, including nanotechnologically derived foods. The European Food Safety Authority (EFSA) provides guidance on risk assessment and safety evaluation of novel foods, including nanomaterials. These regulations and guidelines help ensure that nanotechnologically derived foods meet safety standards before they enter the market.

7.4.3 COST CONSIDERATIONS AND ECONOMIC VIABILITY

Cost considerations play a significant role in the adoption of nanotechnologically derived food production on a large scale. Developing cost-effective manufacturing processes is crucial to ensure economic viability and competitive pricing of these products. Several factors need to be considered to address cost challenges effectively. Efficiency in raw material utilization is essential to minimize costs. This may involve exploring alternative sources of raw materials or developing cost-effective methods for synthesizing nanomaterials. For example, researchers and industry stakeholders can investigate the use of agricultural by-products or waste materials as sources of nanomaterials. By repurposing these materials, costs associated with sourcing new raw materials can be reduced, and waste can be minimized. Optimizing production processes to reduce waste, energy consumption, and water usage can also contribute to cost savings. Implementing process improvements, such as automation, advanced control systems, and continuous manufacturing techniques, can increase efficiency and reduce production costs. Additionally, incorporating sustainable practices in the production processes, such as using energy-efficient equipment or implementing recycling and reusing systems, can lead to cost savings and environmental benefits.

Investments in research and development are necessary to improve process efficiency, reduce production costs, and develop innovative applications of nanotechnology in the food industry. Governments, funding agencies, and industry collaborations can support these efforts by providing financial resources and incentives for research and development activities. By investing in research and development, new technologies and techniques can be developed to optimize production processes and reduce costs associated with nanotechnologically derived food production. Scale economies can also play a role in cost reduction. As production volumes increase, economies of scale may be realized, leading to lower production costs. However, achieving economies of scale requires significant upfront investments and market demand to support large-scale production (Martín-Lara et al., 2020; Menezes et al., 2016; Tavares et al., 2022). Therefore, it is crucial to carefully assess market demand and ensure that there is sufficient consumer acceptance and demand for nanotechnologically derived food products before scaling up production. Long-term sustainability considerations are vital for cost-effectiveness. Life cycle assessments that evaluate the environmental impact of nanotechnologically derived food production can identify areas for improvement and cost reduction. By identifying and addressing potential environmental inefficiencies in the production process, companies can reduce costs associated with waste disposal, energy consumption, and raw material usage. Implementing sustainable practices, such as using renewable energy sources, reducing carbon emissions, and optimizing waste management, can contribute to both cost savings and environmental benefits. Consumer acceptance and willingness to pay for nanotechnologically derived foods are critical for their economic viability. Market research, consumer surveys, and targeted marketing strategies can help gauge consumer preferences and price sensitivity. Understanding consumer perceptions, concerns, and preferences regarding nanotechnology in food can help companies develop pricing strategies that align with consumer expectations. Communicating the benefits and value proposition of these products effectively can enhance consumer perception and willingness to pay a premium. Global examples illustrate efforts to address cost considerations in nanotechnologically derived food production. For instance, in Australia, the Commonwealth Scientific and Industrial Research Organisation (CSIRO) has been working on cost-effective methods for producing nanocellulose, a nanomaterial with various applications in the food industry. Their research focuses on utilizing agricultural waste products, such as sugarcane bagasse, to produce nanocellulose, thus reducing raw material costs and waste.

7.5 APPROACHES FOR RISK EVALUATION AND THE CONTROL OF FOOD-RELATED NANOTECHNOLOGIES

Approaches for risk evaluation and the control of food-related nanotechnologies are essential to ensure the safety and responsible use of nanotechnologically derived food products. Risk assessment and risk management are two key components of the overall process of evaluating and addressing the potential risks associated with the use of nanomaterials in the food industry.

7.5.1 RISK EVALUATION

- **Hazard Identification:** The first step in risk evaluation is to identify the potential hazards associated with nanomaterials used in food products. This involves understanding the physicochemical properties of nanomaterials and their potential interactions with biological systems.
- **Exposure Assessment:** Assessing the exposure to nanomaterials involves understanding the pathways by which consumers may come into contact with these materials and estimating the levels of exposure. This includes evaluating the potential for ingestion, inhalation, or dermal exposure during the production, handling, and consumption of nanotechnologically derived foods.
- **Toxicological Evaluation:** Evaluating the toxicological properties of nanomaterials is crucial to understand their potential health effects. This includes assessing their bioavailability, distribution, metabolism, and potential toxicity. *In vitro* and *in vivo* studies are conducted to evaluate the biological responses and potential adverse effects of nanomaterials.
- **Risk Characterization:** Risk characterization involves integrating the information from hazard identification, exposure assessment, and toxicological evaluation to estimate the level of risk associated with the use of nanomaterials in food products. This step helps determine the potential hazards, their likelihood of occurrence, and the severity of their potential effects on human health.

7.5.2 RISK CONTROL

- **Risk Management Strategies:** Once the risks are characterized, risk management strategies are implemented to control and mitigate potential hazards. This includes establishing regulatory frameworks, guidelines, and standards for the use of nanomaterials in food production. These regulations ensure that manufacturers adhere to safety requirements and provide transparency and accountability in the use of nanotechnologies.

- **Safety-by-Design Approach:** Implementing a safety-by-design approach involves considering safety aspects throughout the entire lifecycle of nanotechnologically derived food products. This includes designing nanomaterials and processes with safety in mind, minimizing potential hazards, and incorporating risk control measures during product development and manufacturing.
- **Labeling and Traceability:** Clear and accurate labeling of nanotechnologically derived food products is essential to provide consumers with information about their composition and the presence of nanomaterials. Labeling requirements enable consumers to make informed choices and contribute to transparency in the marketplace. Additionally, establishing traceability systems helps track the use of nanomaterials throughout the supply chain, facilitating product recalls and ensuring accountability.
- **Post-Market Surveillance:** Ongoing monitoring and post-market surveillance systems are crucial to detect any potential adverse effects or emerging risks associated with nanotechnologically derived food products. Regular inspections, product testing, and data collection help identify and address any safety concerns that may arise after the products are already in the market.
- **Collaboration and Knowledge Sharing:** Collaboration among regulatory agencies, industry stakeholders, scientific communities, and consumer organizations is essential for effective risk evaluation and control. Sharing knowledge, expertise, and data fosters a better understanding of nanotechnology's potential risks and enables the development of robust risk management strategies.

7.6 FUTURE PERSPECTIVES

7.6.1 ENHANCING FOOD QUALITY AND NUTRITION THROUGH NANOENCAPSULATION

Nanoencapsulation is a promising approach that involves enclosing active compounds, such as vitamins, antioxidants, flavors, or bioactive peptides, within nanoscale capsules or particles. This technique offers several advantages in enhancing food quality and nutrition. Nanoencapsulation can protect sensitive bioactive compounds from degradation, oxidation, or interactions with other food components, thereby preserving their functionality and bioavailability. It can also improve the stability and solubility of these compounds, enabling their controlled release and targeted delivery within the body. One application of nanoencapsulation is in fortifying foods with essential nutrients. By encapsulating vitamins or minerals within nanoparticles, their stability can be improved, and their release can be controlled during digestion. This ensures that the nutrients are protected from environmental factors and released in the appropriate gastrointestinal region for optimal absorption. Nanoencapsulation can also enable the delivery of bioactive compounds with low water solubility, such as curcumin or resveratrol, which exhibit health-promoting properties but are challenging to incorporate into food matrices.

Moreover, nanoencapsulation can enhance the sensory attributes of food products. Encapsulating flavors or aromas within nanoparticles can protect them from degradation during processing and storage, allowing for the development of long-lasting flavors or slow-release aroma profiles. This technology can also be used to mask off-flavors or improve the stability of volatile compounds in foods (Boondaeng et al., 2022). Future advancements in nanoencapsulation techniques will focus on improving the efficiency, scalability, and safety of the process. Researchers are exploring novel encapsulation materials, such as biodegradable polymers or lipids, to ensure the safety and biocompatibility of the encapsulated compounds. They are also investigating techniques to enhance the encapsulation efficiency and reduce production costs, such as spray drying, coacervation, or electrostatic methods. By refining these techniques, nanoencapsulation can become a mainstream approach for enhancing food quality, improving nutrient delivery, and providing innovative sensory experiences (Krochmalny et al., 2022; Li et al., 2020; Montes and Rico, 2020).

7.6.2 REAL-TIME MONITORING OF FOOD QUALITY WITH NANOSENSORS

Ensuring the safety and quality of food products is a significant concern for the food industry. Traditional methods of monitoring food quality rely on time-consuming and labor-intensive laboratory analyses, which may not provide real-time information. Nanosensors, on the other hand, offer the potential for rapid, sensitive, and on-site monitoring of food quality parameters. Nanosensors are devices that incorporate nanomaterials to detect and measure specific compounds or properties in food. They can be designed to detect a wide range of analytes, including pathogens, toxins, contaminants, freshness indicators, or changes in pH, temperature, or gas composition. Nanosensors can provide real-time data, enabling timely interventions to maintain food quality and safety.

One example of nanosensor application is the detection of food-borne pathogens. Nanomaterials, such as quantum dots or carbon nanotubes, can be functionalized to selectively bind to specific pathogens, resulting in a measurable signal. This allows for the rapid and sensitive detection of bacterial or viral contamination in food samples. Nanosensors can also be used to monitor the freshness of perishable foods by detecting changes in volatile compounds or enzymatic activity associated with spoilage. The development of nanosensors for food quality monitoring is still in its early stages, but significant progress has been made. Researchers are exploring various nanomaterials, such as nanoparticles, nanowires, or nanofibers, and their integration into sensor platforms. They are also investigating novel signal transduction mechanisms, such as fluorescence, electrochemical, or surface plasmon resonance, to enhance the sensitivity and selectivity of the sensors. Additionally, efforts are being made to ensure the compatibility of nanosensors with food matrices, including addressing issues related to stability, interference, or regulatory compliance. As nanosensors continue to evolve, they hold great promise for revolutionizing food quality monitoring. Their real-time capabilities, portability, and ease of use can enable rapid and informed decision-making throughout the food supply chain, from production to distribution and retail. However, challenges related to standardization, calibration, and cost-effectiveness must be addressed to facilitate their widespread adoption and integration into existing quality control systems.

7.6.3 ADVANCED FOOD PACKAGING MATERIALS FOR EXTENDED SHELF LIFE

Food packaging plays a critical role in preserving the quality and extending the shelf life of food products. Advanced packaging materials incorporating nanotechnology offer innovative solutions for enhancing food safety, maintaining freshness, and reducing food waste. Nanotechnology-enabled packaging materials provide several functionalities that conventional packaging cannot achieve. One such functionality is antimicrobial properties. Nanomaterials, such as silver nanoparticles or nanocomposites, can be incorporated into packaging films or coatings to inhibit the growth of spoilage microorganisms or food-borne pathogens. These antimicrobial properties can significantly reduce the risk of food contamination and extend the shelf life of perishable products. Additionally, nanotechnology can improve the barrier properties of packaging materials. Nanoscale coatings or laminates can create a barrier against oxygen, moisture, or light, which are major factors contributing to food deterioration. By preventing the entry of oxygen or moisture and blocking harmful UV light, nanotechnology-enabled packaging materials can maintain the freshness and quality of food products over an extended period. This can reduce the need for preservatives or artificial additives, contributing to cleaner labels and healthier food options. Furthermore, nanosensors integrated into packaging materials can provide real-time information on the condition of packaged food. These sensors can detect changes in temperature, gas composition, or pathogen presence, providing valuable data on the product's safety and quality. This real-time monitoring capability enables more accurate shelf-life predictions and allows for timely interventions to prevent spoilage or deterioration.

The development and adoption of advanced packaging materials incorporating nanotechnology face challenges related to safety, regulatory compliance, and scalability. The safety of nanomaterials used in packaging must be thoroughly evaluated to ensure their inertness, stability, and absence of migration into food. Regulatory frameworks need to be established to address the specific requirements and potential risks associated with nanotechnology-enabled packaging materials. Additionally, the scalability and cost-effectiveness of manufacturing processes must be considered to facilitate their widespread adoption by the food industry.

7.7 CONCLUSION

7.7.1 SUMMARY OF CHALLENGES

The integration of nanotechnology into large-scale food production comes with several challenges that need to be addressed for its successful adoption. Scaling up nanotechnology processes requires the development of efficient and cost-effective manufacturing methods that can produce nanomaterials in large quantities while maintaining consistent quality. This involves process optimization, selection and sourcing of raw materials, and establishing stable supply chains. Collaboration between researchers, industry stakeholders, and technology providers is crucial to overcome these challenges and accelerate the development of scalable nanotechnology processes. Maintaining safety and quality is another critical aspect of scaling up nanotechnology processes. Thorough evaluation of the safety and regulatory aspects associated with nanomaterials is essential to ensure that the scaled-up production processes do not compromise the safety of the final food products. Characterization and risk assessment of nanomaterials play a pivotal role in maintaining safety, requiring reliable and standardized methods to assess their properties, behavior, and potential risks. Implementing quality control and assurance processes, along with robust quality management systems, ensures that safety and quality are maintained throughout the production process.

Cost considerations are significant in the adoption of nanotechnologically derived food production on a large scale. Developing cost-effective manufacturing processes involves optimizing raw material utilization, reducing waste and energy consumption, and investing in research and development. Long-term sustainability considerations, such as life cycle assessments and implementing sustainable practices, also contribute to cost-effectiveness. Consumer acceptance and willingness to pay for nanotechnologically derived foods are critical for their economic viability, requiring market research, effective communication, and targeted marketing strategies.

7.7.2 FUTURE PROSPECTS AND OPPORTUNITIES

Despite the challenges, the future prospects and opportunities of nanotechnology in the food industry are highly promising. The potential benefits of nanotechnology in enhancing food quality, safety, and nutrition are vast. Nanoencapsulation techniques offer the ability to protect sensitive bioactive compounds, fortify foods with essential nutrients, and improve sensory attributes. Continued research and development efforts are expected to enhance the efficiency, scalability, and safety of nanoencapsulation processes, making them more accessible for large-scale food production. Real-time monitoring of food quality using nanosensors presents a significant opportunity for rapid and on-site detection of contaminants, freshness indicators, and changes in food parameters. Advancements in nanomaterials, sensor integration, and signal transduction mechanisms will contribute to the development of more sensitive and selective nanosensors. Integration of nanosensors into existing quality control systems can enable timely interventions and improve overall food safety and quality management.

Advanced food packaging materials incorporating nanotechnology offer extended shelf life, improved barrier properties, and real-time monitoring capabilities. Antimicrobial properties, enhanced barrier performance, and integrated nanosensors can contribute to reducing food waste and ensuring the freshness and safety of food products. Overcoming challenges related to safety, regulatory compliance, scalability, and cost-effectiveness will pave the way for the widespread adoption of these advanced packaging materials. The future of nanotechnology in the food industry also relies on collaboration between regulatory agencies, industry stakeholders, and scientific communities. Comprehensive safety guidelines and regulatory frameworks need to be developed to address the unique challenges associated with scaling up nanotechnology processes and ensure the safe production and use of nanotechnologically derived foods. Government support, funding agencies, and industry collaborations play a crucial role in providing the necessary resources, incentives, and research funding to drive innovation and accelerate the integration of nanotechnology into large-scale food production. In conclusion, the scalability and cost-effectiveness of nanotechnology processes, along with maintaining safety and quality, are critical considerations for the widespread adoption of nanotechnologically derived food production. Despite the challenges, the future prospects and opportunities of nanotechnology in the food industry are immense. Continued research, collaboration, and investment in developing scalable processes, ensuring safety and regulatory compliance, and addressing cost considerations will enable the food industry to harness the full potential of nanotechnology. By doing so, we can enhance food quality, safety, and nutrition, reduce food waste, and contribute to a more sustainable and efficient food supply chain.

KEYWORDS

- **agriculture**
- **food**
- **challenges**
- **environmental sustainability**
- **nanotechnology**
- **packaging**
- **safety**

REFERENCES

Alzeer, J., & Abou Hadeed, K. (2016). Ethanol and its Halal status in food industries, *Trends in Food Science and Technology, 58*, 14–20. https://doi.org/10.1016/j.tifs.2016.10.018.

Angelopoulou, P., Giaouris, E., & Gardikis, K. (2022). Applications and prospects of nanotechnology in food and cosmetics preservation. *Nanomaterials, 12*, 1–28. https://doi.org/10.3390/nano12071196.

Asimakopoulos, G., Baikousi, M., Salmas, C., Bourlinos, A. B., Zboril, R., & Karakassides, M. A. (2021). Advanced Cr(VI) sorption properties of activated carbon produced via pyrolysis of the "Posidonia oceanica" seagrass. *Journal of Hazardous Materials, 405* (2021) 124274. https://doi.org/10.1016/j.jhazmat.2020.124274.

Babu, P. J. (2022). Nanotechnology mediated intelligent and improved food packaging. *International Nano Letters, 12*, 1–14. https://doi.org/10.1007/s40089-021-00348-8.

Baghi, F., Gharsallaoui, A., Dumas, E., & Ghnimi, S. (2022). Advancements in biodegradable active films for food packaging: Effects of nano/microcapsule incorporation. *Foods 11*, 1–44. https://doi.org/10.3390/foods11050760.

Bashir, S., Thakur, A., Lgaz, H., Chung, I. M., & Kumar, A. (2020). Corrosion inhibition efficiency of bronopol on aluminium in 0.5 MHCl solution: Insights from experimental and quantum chemical studies. *Surfaces and Interfaces, 20*, 100542. https://doi.org/10.1016/j.surfin.2020.100542.

Bashir, S., Thakur, A., Lgaz, H., Chung, I.-M., & Kumar, A. (2019). Computational and experimental studies on Phenylephrine as anti-corrosion substance of mild steel in acidic medium. *Journal of Molecular Liquids, 293*, 111539. https://doi.org/10.1016/j.molliq.2019.111539.

Boondaeng, A., Khanoonkon, N., Vaithanomsat, P., Apiwatanapiwat, W., Trakunjae, C., Janchai, P., & Niyomvong, N. (2022). Recovery and purification of fumaric acid from fermented oil palm empty fruit bunches using a simple two-stage precipitation procedure. *Fermentation, 8*. https://doi.org/10.3390/fermentation8030121.

Cai, L., Rice, S., Koziel, J. A., Jenks, W. S., & van Leeuwen, J. H. (2016). Further purification of food-grade alcohol to make a congener-free product, *Journal of the Institute of Brewing, 122*, 84–92. https://doi.org/10.1002/jib.295.

Calabrese, M. G., & Ferranti, P. (2018). Novel foods: New food sources. *Encyclopedia of Food Security and Sustainability,* 271–275. https://doi.org/10.1016/B978-0-08-100596-5.22128-8.

Canteli, A. M. D., Carpiné, D., de P. Scheer, A., Mafra, M. R., & Igarashi-Mafra, L. (2014). Fixed-bed column adsorption of the coffee aroma compound benzaldehyde from aqueous solution onto granular activated carbon from coconut husk. *LWT—Food Science and Technology, 59*, 1025–1032. https://doi.org/10.1016/j.lwt.2014.06.015.

Cardoso-Ugarte, G. A., López-Malo, A., & Jiménez-Munguía, M. T. (2016). *Application of nanoemulsion technology for encapsulation and release of lipophilic bioactive compounds in food*. Elsevier Inc. https://doi.org/10.1016/b978-0-12-804306-6.00007-6.

Chau, C. F., Wu, S. H., & Yen, G. C. (2007). The development of regulations for food nanotechnology. *Trends in Food Science and Technology, 18*, 269–280. https://doi.org/10.1016/j.tifs.2007.01.007.

Chausali, N., Saxena, J., & Prasad, R. (2022). Recent trends in nanotechnology applications of bio-based packaging. *Journal of Agriculture and Food Research, 7*, 100257. https://doi.org/10.1016/j.jafr.2021.100257.

Fadiji, A. E., Mthiyane, D. M. N., Onwudiwe, D. C., & Babalola, O. O. (2022). Harnessing the known and unknown impact of nanotechnology on enhancing food security and reducing postharvest losses: Constraints and future prospects. *Agronomy, 12*. https://doi.org/10.3390/agronomy12071657.

Fayaz, M., Sajad Rabani, M., Ahmad Wani, S., & Ahmad Thoker, S. (2020). Nano-agriculture: A novel approach in agriculture, microbiota biofertilizers a sustain. *Continuum for Plant and Soil Health* (pp. 99–122). https://doi.org/10.1007/978-3-030-48771-3_7.

Ganjoo, R., Sharma, S., Thakur, A., & Kumar, A. (2022). Thermodynamic study of corrosion inhibition of dioctylsulfosuccinate sodium salt as corrosion inhibitor against mild steel in 1 M HCl. *Materials Today: Proceedings,* 1–5. https://doi.org/10.1016/j.matpr.2022.05.594.

Golovacheva, N. E., Morozova, S. S., Abramova, I. M., & Mukhin, V. M. (2021). Active coals as an important factor in improving the efficiency of the production of alcoholic beverages. *IOP Conference Series: Earth and Environmental Science, 839*. https://doi.org/10.1088/1755-1315/839/2/022052.

Granato, D., de Oliveira, C. C., Caruso, M. S. F., Nagato, L. A. F., & Alaburda, J. (2014). Feasibility of different chemometric techniques to differentiate commercial Brazilian sugarcane spirits based on chemical markers. *Food Research International, 60*, 212–217. https://doi.org/10.1016/j.foodres.2013.09.044.

Gutiérrez, T. J., Ponce, A. G., & Alvarez, V. A. (2017). Nano-clays from natural and modified montmorillonite with and without added blueberry extract for active and intelligent food nanopackaging materials. *Materials Chemistry and Physics, 194*, 283–292. https://doi.org/10.1016/j.matchemphys.2017.03.052.

Hosseini, H., & Jafari, S. M. (2020). Introducing nano/microencapsulated bioactive ingredients for extending the shelf-life of food products., *Advances in Colloid and Interface Science, 282*, 102210. https://doi.org/10.1016/j.cis.2020.102210.

Hou, T., Sankar Sana, S., Li, H., Wang, X., Wang, Q., Boya, V. K. N., Vadde, R., Kumar, R., Kumbhakar, D. V., Zhang, Z., & Mamidi, N. (2022). Development of plant protein derived tri angular shaped nano zinc oxide particles with inherent antibacterial and neurotoxicity properties. *Pharmaceutics, 14*. https://doi.org/10.3390/pharmaceutics14102155.

Jafarzadeh, S., Salehabadi, A., & Jafari, S. M. (2020). Metal nanoparticles as antimicrobial agents in food packaging. INC. https://doi.org/10.1016/B978-0-12-815866-1.00010-8.

Jagtiani, E. (2022). Advancements in nanotechnology for food science and industry. *Food Frontiers, 3*, 56–82. https://doi.org/10.1002/fft2.104.

Jain, A., Ranjan, S., Dasgupta, N., & Ramalingam, C. (2018). Nanomaterials in food and agriculture: An overview on their safety concerns and regulatory issues. *Critical Reviews in Food Science and Nutrition, 58*, 297–317. https://doi.org/10.1080/10408398.2016.1160363.

Kaur, H., Thakur, R. C., & Kumar, H. (2020). Effect of proteinogenic amino acids L-serine/L-threonine on volumetric and acoustic behavior of aqueous 1-butyl-3-propyl imidazolium bromide at T = (288.15, 298.15, 308.15, 318.15) K. *Journal of Chemical Thermodynamics, 150*, 106211. https://doi.org/10.1016/j.jct.2020.106211.

Kaur, H., Thakur, R. C., Kumar, H., & Katal, A. (2021). Effect of α–amino acids (glycine, L-alanine, L-valine and L-leucine) on volumetric and acoustic properties of aqueous 1-Butyl-3-propylimidazolium bromide at T = (288.15, 298.15, 308.15, 318.15) K. *Journal of Chemical Thermodynamics, 158*, 106433. https://doi.org/10.1016/j.jct.2021.106433.

Kaur, H., Thakur, R. C., Pathania, V. S., & Sharma, S. (2022). Effect of choline-based ionic liquid (*Cholinium ethanoate*) on volumetric and acoustic properties of aliphatic amino acids (glycine and L-alanine). *Journal of Molecular Liquids, 346*, 118247. https://doi.org/10.1016/j.molliq.2021.118247.

Khan, F., Pandey, P., & Upadhyay, T. K. (2022). Applications of nanotechnology-based agrochemicals in food security and sustainable agriculture: An overview. *Agriculture 12*, 1–13. https://doi.org/10.3390/agriculture12101672.

Khezerlou, A., Tavassoli, M., Sani, M. A., Mohammadi, K., Ehsani, A., & McClements, D. J. (2021). Application of nanotechnology to improve the performance of biodegradable biopolymer-based packaging materials. *Polymers (Basel), 13*, 1–21. https://doi.org/10.3390/polym13244399.

Krochmalny, K., Pawlak-Kruczek, H., Skoczylas, N., Kudasik, M., Gajda, A., Gnatowska, R., Serafin-Tkaczuk, M., Czapka, T., Jaiswal, A. K., Vishwajeet, Arora, A., Hardy, T., Jackowski, M., Ostrycharczyk, M., & Niedźwiecki, Ł. (2022). Use of hydrothermal carbonization and cold atmospheric plasma for surface modification of Brewer's spent grain and activated carbon. *Energies, 15*, 4396. https://doi.org/10.3390/en15124396.

Kumar, A., & Thakur, A. (2020). *Encapsulated nanoparticles in organic polymers for corrosion inhibition*. https://doi.org/10.1016/b978-0-12-819359-4.00018-0.

Lade, B. D., & Gogle, D. P. (2019). Nano-biopesticides: Synthesis and applications in plant safety. *Nanotechnology in the Life Sciences,* 169–189. https://doi.org/10.1007/978-3-030-13296-5_9.

Lamri, M., Bhattacharya, T., Boukid, F., Chentir, I., Dib, A. L., Das, D., Djenane, D., & Gagaoua, M. (2021). Meat quality and safety. *Foods, 10* (2021) 2633.

Leonardo, L. M., de O. S. Anna, C., Murilo, S., de P. Nádia, F., Luciana, M. S., & Mariana, C. F. (2017). Post-treatment of cachaca (Brazilian sugarcane spirit) with charcoal made from cane bagasse. *African Journal of Microbiology Research, 11*, 287–295. https://doi.org/10.5897/ajmr2016.8415.

Li, X., Bei, E., Qiu, Y., Xiao, H., Wang, J., Lin, P., Zhang, X., & Chen, C. (2021). Intake of volatile nitrosamines by Chinese residents in different provinces via food and drinking water. *Science of the Total Environment, 754*, 142121. https://doi.org/10.1016/j.scitotenv.2020.142121.

Li, Y., Fang, S., Zhou, X., Zhao, Z., Li, F., & Liu, P. (2020). Adsorption study of lignin removal from recycled alkali black liquor by adsorption resins for improved cellulase hydrolysis of corn straw. *Molecules, 25*, 1–16. https://doi.org/10.3390/molecules25194475.

Liskova, A., Samec, M., Koklesova, L., Brockmueller, A., Zhai, K., Abdellatif, B., Siddiqui, M., Biringer, K., Kudela, E., Pec, M., Gadanec, L. K., Šudomová, M., Hassan, S. T. S., Zulli, A., Shakibaei, M., Giordano, F. A., Büsselberg, D., Golubnitschaja, O., & Kubatka, P. (2021). Flavonoids as an effective sensitizer for anti-cancer therapy: Insights into multi-faceted mechanisms and applicability towards individualized patient profiles. *EPMA Journal, 12*, 155–176. https://doi.org/10.1007/s13167-021-00242-5.

Lopes, G. K. P., Zanella, H. G., Spessato, L., Ronix, A., Viero, P., Fonseca, J. M., Yokoyama, J. T. C., Cazetta, A. L., & Almeida, V. C. (2021). Steam-activated carbon from malt bagasse: Optimization of preparation conditions and adsorption studies of sunset yellow food dye. *Arabian Journal of Chemistry, 14*. https://doi.org/10.1016/j.arabjc.2021.103001.

Luo, Y., Kong, L., Xue, R., Wang, W., & Xia, X. (2020). Bitterness in alcoholic beverages: The profiles of perception, constituents, and contributors. *Trends in Food Science and Technology, 96*, 222–232. https://doi.org/10.1016/j.tifs.2019.12.026.

Martín-Lara, M. A., Calero, M., Ronda, A., Iáñez-Rodríguez, I., Escudero, C. (2020). Adsorptive behavior of an activated carbon for bisphenol A removal in single and binary (bisphenol A-heavy metal) solutions. *Water (Switzerland), 12*. https://doi.org/10.3390/W12082150.

Marynchenko, L., Marynchenko, V., & Hyvel, M. (2017). Research of mineral adsorbents application for water-alcohol solutions purification in technology of alcoholic beverages. *EUREKA: Physics and Engineering, 4*, 3–10. https://doi.org/10.21303/2461-4262.2017.00397.

Menezes, A. G. T., Menezes, E. G. T., Alves, J. G. L. F., Rodrigues, L. F., & das G. Cardoso, M. (2016). Vodka production from potato (*Solanum tuberosum* L.) using three Saccharomyces cerevisiae isolates. *Journal of the Institute of Brewing, 122*, 76–83. https://doi.org/10.1002/jib.302.

Mohammad, Z. H., Ahmad, F., Ibrahim, S. A., & Zaidi, S. (2022). Application of nanotechnology in different aspects of the food industry. *Discover Food, 2*. https://doi.org/10.1007/s44187-022-00013-9.

Montes, J. A., & Rico, C. (2020). Biogas potential of wastes and by-products of the alcoholic beverage production industries in the Spanish region of cantabria. *Applied Sciences, 10*, 1–15. https://doi.org/10.3390/app10217481.

Nidhi, H. K. (2022). Advancement in field of energy storage applications by using ionic liquids. *Journal of Physics: Conference Series, 2267*. https://doi.org/10.1088/1742-6596/2267/1/012043.

Ogunkalu, O. A. (2019). Utilization of nanotechnology in aquaculture and seafood sectors. *Eurasian Journal of Food Science and Technology, 3*, 26–33.

Oliveira, P. R., Lamy-Mendes, A. C., Rezende, E. I. P., Mangrich, A. S., Marcolino Junior, L. H., & Bergamini, M. F. (2015). Electrochemical determination of copper ions in spirit drinks using carbon paste electrode modified with biochar. *Food Chemistry, 171*, 426–431. https://doi.org/10.1016/j.foodchem.2014.09.023.

Onuki, S., Koziel, J. A., Jenks, W. S., Cai, L., Grewell, D., & van Leeuwen, J. H. (2016). Taking ethanol quality beyond fuel grade: A review, *Journal of the Institute of Brewing, 122*, 588–598. https://doi.org/10.1002/jib.364.

Onuki, S., Koziel, J. A., Jenks, W. S., Cai, L., Rice, S., & Van Leeuwen, J. (2015). Ethanol purification with ozonation, activated carbon adsorption, and gas stripping, *Separation and Purification Technology, 151*, 165–171. https://doi.org/10.1016/j.seppur.2015.07.026.

Panpatte, D. G., & Jhala, Y. K. (2019). Nanotechnology for agriculture: Crop production & protection. https://doi.org/10.1007/978-981-32-9374-8.

Paramo, L. A., Feregrino-Pérez, A. A., Guevara, R., Mendoza, S., & Esquivel, K. (2020). Nanoparticles in agroindustry: Applications, toxicity, challenges, and trends. *Nanomaterials, 10*, 1–33. https://doi.org/10.3390/nano10091654.

Parveen, G., Bashir, S., Thakur, A., Saha, S. K., Banerjee, P., & Kumar, A. (2020). Experimental and computational studies of imidazolium based ionic liquid 1-methyl- 3-propylimidazolium iodide on mild steel corrosion in acidic solution Experimental and computational studies of imidazolium based ionic liquid 1-methyl-3-propylimidazolium. *Materials Research Express, 7*, 016510. https://doi.org/https://doi.org/10.1088/2053-1591/ab5c6a.

Pauley, M., & Maskell, D. (2017). Mini-review: The role of saccharomyces cerevisiae in the production of gin and vodka. *Beverages, 3*, 1–11. https://doi.org/10.3390/beverages3010013.

Prasad, R. D., Sahoo, A. K., Shrivastav, O. P., Charmode, N., Prasad, S. R., Kamat, R., Kajave, N. G., Chauhan, J., Banga, S., Tamboli, U., Pandharpatte, M. S., Atigre, R. H., Shaikh, V., Padvi, M. N., Salvalkar, P., & Prasad, N. R. (2022). ES food and agroforestry a review on aspects of nanotechnology in food science and animal nutrition. *ES Food Agroforestry, 8*, 12–46. https://dx.doi.org/10.30919/esfaf704.

Prasad, R., Kumar, V., & Kumar, M (2017). Nanotechnology: Food and environmental paradigm. (pp. 1–344). https://doi.org/10.1007/978-981-10-4678-0.

Primoži˘c, M., Knez, Ž. , & Leitgeb, M. (Bio). (2021). Nanotechnology in food science—Food packaging. *Nanomaterials, 11*, 292.

Rashidi, L., & Khosravi-Darani, K. (2011). The applications of nanotechnology in food industry, *Critical Reviews in Food Science and Nutrition, 51*, 723–730. https://doi.org/10.1080/10408391003785417.

Raspor, P., & Goranovič, D. (2008). Biotechnological applications of acetic acid bacteria, *Critical Reviews in Biotechnology, 28*, 101–124. https://doi.org/10.1080/07388550802046749.

Sahoo, M., Panigrahi, C., Vishwakarma, S., & Kumar, J. (2022). A review on nanotechnology: Applications in food industry, future opportunities, challenges and potential risks. *Journal of Nanotechnology and Nanomaterials, 3* (2022) 28–33. https://doi.org/10.33696/nanotechnol.3.029.

Sandilya, S., Thakur, A., Suresh Singh, S., & Kumar, A. (2021). Recent advances , synthesis and characterization of bio-based based polymers from natural sources. *Plant Cell Biotechnology and Molecular Biology, 22*, 70–93.

Shafiq, M., Anjum, S., Hano, C., Anjum, I., & Abbasi, B. H. (2020). An overview of the applications of nanomaterials and nanodevices in the food industry. *Foods, 9*, 1–27. https://doi.org/10.3390/foods9020148.

Sharma, D., et al. (2014). Synthesis, electrochemical, morphological, computational and corrosion inhibition studies of 3-(5-naphthalen-2-yl-[1,3,4]oxadiazol-2-yl)-pyridine against mild steel in 1MHCl. *Asian Journal of Chemistry, 35*, 1079–1088. https://doi.org/https://doi.org/10.14233/ajchem.2023.27711.

Shibata, J., Murayama, N., & Tateyama, M. (2009). Adsorption of purine compounds in beer with activated carbon prepared from beer lees. *Resource Processes, 56*, 120–126. https://doi.org/10.4144/rpsj.56.120.

Siegrist, M., Cousin, M. E., Kastenholz, H., & Wiek, A. (2007). Public acceptance of nanotechnology foods and food packaging: The influence of affect and trust. *Appetite, 49* (2007) 459–466. https://doi.org/10.1016/j.appet.2007.03.002.

Siegrist, M., Stampfli, N., Kastenholz, H., & Keller, C. (2008) Perceived risks and perceived benefits of different nanotechnology foods and nanotechnology food packaging. *Appetite, 51*, 283–290. https://doi.org/10.1016/j.appet.2008.02.020.

Simate, G. S., Iyuke, S. E., Ndlovu, S., Heydenrych, M., & Walubita, L. F. (2012). Human health effects of residual carbon nanotubes and traditional water treatment chemicals in drinking water. *Environment International, 39*, 38–49. https://doi.org/10.1016/j.envint.2011.09.006.

Singh Sekhon, B. (2014). Nanotechnology in agri-food production: An overview. *Nanotechnology, Science and Applications, 7*, 31–53. https://doi.org/10.2147/NSA.S39406.

Siríštová, L., Prinosilová, S., Riddellová, K., Hajšlová, J., & Melzoch, K. (2012). Changes in quality parameters of vodka filtered through activated charcoal. *Czech Journal of Food Sciences, 30*, 474–482. https://doi.org/10.17221/361/2011-cjfs.

Soukoulis, C., & Bohn, T. (2018). A comprehensive overview on the micro- and nano-technological encapsulation advances for enhancing the chemical stability and bioavailability of carotenoids. *Critical Reviews in Food Science and Nutrition, 58*, 1–36. https://doi.org/10.1080/10408398.2014.971353.

Sozer, N., & Kokini, J. L. (2009). Nanotechnology and its applications in the food sector. *Trends in Biotechnology,* 27 (2009) 82–89. https://doi.org/10.1016/j.tibtech.2008.10.010.

Tavares, A. P. M., Gonçalves, M. J. A., Brás, T., Pesce, G. R., Xavier, A. M. R. B., & Fernandes, M. C. (2022). Cardoon hydrolysate detoxification by activated carbon or membranes system for bioethanol production. *Energies, 15*. https://doi.org/10.3390/en15061993.

Thakur, A., & Kumar, A. (2021). Sustainable inhibitors for corrosion mitigation in aggressive corrosive media: a comprehensive study. *Journal of Bio- and Tribo-Corrosion, 7*, 1–48. https://doi.org/10.1007/s40735-021-00501-y.

Thakur, A., & Kumar, A. (2022). Recent advances on rapid detection and remediation of environmental pollutants utilizing nanomaterials-based (bio)sensors. *Science of the Total Environment, 834*, 155219. https://doi.org/10.1016/j.scitotenv.2022.155219.

Thakur, A., & Kumar, A. (2023). Recent trends in nanostructured carbon-based electrochemical sensors for the detection and remediation of persistent toxic substances in real- time analysis. *Materials Research Express, 10*, 034001. https://doi.org/https://doi.org/10.1088/2053-1591/acbd1a.

Thakur, A., Kaya, S., & Kumar, A. (2021). Recent innovations in nano container-based self-healing coatings in the construction industry. *Current Nanoscience, 18*, 203–216. https://doi.org/10.2174/1573413717666210216120741.

Thakur, A., Kaya, S., Abousalem, A. S., & Kumar, A. (2022a). Experimental , DFT and MC simulation analysis of Vicia Sativa weed aerial extract as sustainable and eco-benign corrosion inhibitor for mild steel in acidic environment. *Sustainable Chemistry and Pharmacy. 29*, 100785. https://doi.org/10.1016/j.scp.2022.100785.

Thakur, A., Kaya, S., Abousalem, A. S., Sharma, S., Ganjoo, R., Assad, H., & Kumar, A. (2022b). Computational and experimental studies on the corrosion inhibition performance of an aerial extract of Cnicus Benedictus weed on the acidic corrosion of mild steel, *Process Safety and Environmental Protection, 161*, 801–818. https://doi.org/10.1016/j.psep.2022.03.082.

Thakur, A., Kumar, A., Kaya, S., Marzouki, R., Zhang, F., & Guo, L. (2022c). Recent advancements in surface modification, characterization and functionalization for enhancing the biocompatibility and corrosion resistance of biomedical implants. *Coatings, 12*, 1459. https://doi.org/10.3390/coatings12101459.

Thakur, A., Kumar, A., Kaya, S., Vo, D. V. N., & Sharma, A. (2022d). Suppressing inhibitory compounds by nanomaterials for highly efficient biofuel production: A review. *Fuel, 312*, 122934. https://doi.org/10.1016/j.fuel.2021.122934.

Thakur, A., Kumar, A., Sharma, S., Ganjoo, R., & Assad, H. (2022e). Computational and experimental studies on the efficiency of Sonchus arvensis as green corrosion inhibitor for mild steel in 0.5 M HCl solution. *Materials Today: Proceedings, 66*, 609–621. https://doi.org/10.1016/j.matpr.2022.06.479.

Thakur, A., Sharma, S., Ganjoo, R., Assad, H., & Kumar, A. (2022f). Anti-corrosive potential of the sustainable corrosion inhibitors based on biomass waste: A review on preceding and perspective research. *Journal of Physics: Conference Series,* 2267 (2022) 012079. https://doi.org/10.1088/1742-6596/2267/1/012079.

Thakur, A., Kumar, A., & Zhang, R. (2023). Alcoholic beverage purification applications of activated carbon. In C. Verma, & M. A. Quraishi (Eds.). *Activated Carbon: Progress and Applications.* The Royal Society of Chemistry. https://doi.org/10.1039/BK9781839169861-00152.

Thakur, R. C., Kaur, R., & Kaur, H. (2020). Transport properties of thiaminehydrochlorideinbinary aqueous mixtures of galactose. *European Journal of Molecular and Clinical Medicine, 7*, 2020.

Valdés, M. G., González, A. C. V., Calzón, J. A. G., & Díaz-García, M. E. (2009). Analytical nanotechnology for food analysis. *Microchimica Acta, 166*, 1–19. https://doi.org/10.1007/s00604-009-0165-z.

Verma, C., Thakur, A., Ganjoo, R., Sharma, S., & Assad, H. (2023). Coordination bonding and corrosion inhibition potential of nitrogen-rich heterocycles : Azoles and triazines as specific examples. *Coordination Chemistry Reviews, 488*, 215177. https://doi.org/10.1016/j.ccr.2023.215177.

Vijayakumar, M. D., Surendhar, G. J., Natrayan, L., Patil, P. P., Ram, P. M. B., & Paramasivam, P. (2022). Evolution and recent scenario of nanotechnology in agriculture and food industries. *Journal of Nanomaterials, 2022*. https://doi.org/10.1155/2022/1280411.

Weiss, J., Takhistov, P., & McClements, D. J. (2006). Functional materials in food nanotechnology, *Journal of Food Science, 71*, 107–116. https://doi.org/10.1111/j.1750-3841.2006.00195.x.

Wu, J., Liu, Y., Zhao, H., Huang, M., Sun, Y., Zhang, J., & Sun, B. (2021). Recent advances in the understanding of off-flavors in alcoholic beverages: Generation, regulation, and challenges. *Journal of Food Composition and Analysis, 103*, 104117. https://doi.org/10.1016/j.jfca.2021.104117.

Xing, X., Jiang, W., Li, S., Zhang, X., & Wang, W. (2019). Preparation and analysis of straw activated carbon synergetic catalyzed by $ZnCl_2$–H_3PO_4 through hydrothermal carbonization combined with ultrasonic assisted immersion pyrolysis. *Waste Management, 89*, 64–72. https://doi.org/10.1016/j.wasman.2019.04.002.

CHAPTER 8

Nanotechnological Approaches to the Advancement of Innovations in Food Technology

KARTIK SHARMA[1], AADISHA SAINI[2], SATISH KUMAR[3], NEHA GAUTAM[4], and VIKAS KUMAR[2*]

[1]Unit of Innovative Food Packaging and Biomaterials, School of Agro-Industry, Mae Fah Luang University, Chiang Rai, Thailand

[2]ICAR-Indian Institute of Millets Research, Rajendranagar, Hyderabad, Telangana, India

[3]Department of Food Science and Technology, Dr. Y. S. Parmar University of Horticulture and Forestry, Solan, Himachal Pradesh, India

[4]Department of Basic Sciences, Dr. Y. S. Parmar University of Horticulture and Forestry, Nauni, Solan, Himachal Pradesh, India

*Corresponding author

ABSTRACT

This chapter explores the integration of nanotechnology in the food industry, emphasizing its potential to drive advancements in food safety, packaging, and contamination control. Nanotechnology, defined as the manipulation of matter at an atomic and molecular scale (ranging from 1 to 100 nm), holds promise for revolutionizing food technology by offering novel tools and methods for enhancing food quality, extending shelf life, and ensuring food safety. The chapter discusses the various applications of nanomaterials such as nanopackaging and

Nanotechnological Advancements in the Food Industry.
Ishrat Majid, Bababode Adesegun Kehinde, and Sheikh Shafat Hussain (Eds.)

DOI: 10.1201/9781774919750-8

nanocomposites, which can be utilized to prevent contamination and improve the overall sustainability of food production processes. The authors further explore the scientific underpinnings of nanotechnology in food science, highlighting its versatility across fields including mechanics, medicine, and agriculture. The significance of nanotechnology in transforming traditional food systems is underscored, offering a pathway to smarter, safer, and more sustainable food production practices.

8.1 INTRODUCTION TO NANOTECHNOLOGY IN FOOD INDUSTRY

Nanotechnology is the study of atomic and molecular size modification of materials. It can alternatively be described as the manipulation of matter on an atomic and molecular scale, having at least one distinctive dimension measured in nanometers. It has a scale that varies from 1 to 100 nm, having applications in mechanics, medicine as well as in foods, which may be used to develop new tools and methods. In terms of attributes and transport, a nanoparticle is a tiny item that functions as a single entity. They are categorized in accordance with their traits, sizes, and architectures. The field of nanomaterials approaches nanotechnology from the perspective of materials science. Pasteur's introduction of the pasteurization technique, which killed spoilage germs at 1000 nm, set the stage for the development of food nanotechnology and the advancement of food quality. Later, the 2.5 nm model of DNA structure developed by Watson and Crick opened the door to biotechnology, biomedicine, agricultural, and manufacturing process applications. The invention of the 1 nm-sized "buckyball fullerene" carbon nanotube served as a cutting-edge discovery for the innovative globe and established in the era of nanoscience (Chellaram et al., 2014). Besides this, in comparison to their original counterparts, the materials derived from nanotechnology possess a high surface-to-volume ratio along with other novel physiochemical properties, like diffusivity, magnetic, thermodynamic, optical, strength, toxicity, color, solubility, etc. (Gupta et al., 2016). Both industrialized and developing nations are interested in increasing their investments in this technology as nanotechnology has sparked a new economic revolution. As a result, nanotechnology presents a wide range of potential for creation and implementation of systems, materials, or structures with novel qualities in several industries, including food, medicine, agriculture, and others.

8.2 APPLICATIONS OF NANOTECHNOLOGY IN FOOD QUALITY AND SAFETY

Increasing consumer concerns regarding the positive effects and quality of food have prompted researchers to explore methods for enhancing food quality without compromising its nutritional value. As numerous nanoparticle-based products contain essential nutrients and have demonstrated non-toxic properties, their demand has significantly increased in the food industry. In addition, nanotechnology has been found to provide stability to packaging materials even under high pressures and temperatures. It offers comprehensive solutions for the entire food production process, from manufacturing to processing and final packaging. By incorporating nanotechnology, food industry professionals can address various challenges and improve the overall quality and safety of food products. The stability and versatility of nanomaterials make them valuable tools in developing innovative approaches for food manufacturing and packaging. Apart from this, nanomaterials play a significant role in enhancing food quality, ensuring safety, and maximizing the health benefits provided by food. In the field of food science, various businesses, academic institutions, and industries are actively developing innovative strategies, processes, and products that directly utilize nanotechnology. Many organizations, researchers, and industries are actively exploring novel techniques, methods, and products that directly employ nanotechnology in the field of food science. The application of nanotechnology in food science has gained significant attention due to its potential to revolutionize various aspects of food production, processing, and packaging. These endeavors aim to enhance food safety, improve nutritional value, extend shelf life, and develop innovative delivery systems for bioactive compounds. With ongoing research and collaboration, nanotechnology continues to hold great promise for advancing the field of food science (Dasgupta et al., 2015). Nanotechnology applications in the food industry can be broadly divided into two main categories: food nanostructured components and food nanosensing. Food nanostructured components involve the utilization of ingredients with nanostructures, which have various applications in food processing and packaging. These nanostructures can serve as food additives, targeted nutrient carriers, anti-caking agents, antimicrobial agents, and fillers to enhance the strength and durability of packaging materials, among other functions. On the other hand, food nanosensing aims to enhance the assessment of food quality and safety by employing nanoscale sensing technologies (Ezhilarasi et al., 2013).

8.2.1 NANOTECHNOLOGY IN FOOD PROCESSING

Food ingredients with nanostructures are being developed to enhance the consistency, taste, and texture of food products. Nanotechnology has played a significant role in extending the shelf life of various food items and reducing food waste caused by microbial contamination. Nanocarriers are currently utilized to deliver food additives, allowing them to be incorporated into food products without significantly altering their fundamental characteristics. This enables the precise and controlled release of additives, leading to improved product quality and enhanced consumer experience. The size of nanoparticles has been found to influence their absorption by different cell lines, with submicron nanoparticles being more effectively absorbed than larger micro-particles. Thus, particle size plays a role in the distribution of bioactive molecules throughout the body. An ideal delivery system should precisely transport active compounds to the desired location, ensuring availability at the required time and rate, and maintaining optimal levels for extended periods, including during storage. Nanotechnology provides efficient delivery methods through various techniques such as encapsulation, emulsions, biopolymer matrices, simple solutions, and association colloids, all of which possess the desired properties mentioned above. In the field of food packaging, nano polymers are being investigated as alternatives to traditional materials, offering enhanced barrier properties and preservation capabilities. Nanosensors are utilized to detect the presence of pollutants, mycotoxins, and microbes in food, enabling rapid and accurate food safety monitoring. These advancements in nanotechnology hold great potential for improving food quality, safety, and sustainability (Bratovcic et al., 2015). Various applications of nanotechnology in the field of food science are listed in Table 8.1.

TABLE 8.1 Application of Nanotechnology in Food Quality and Safety.

Sl. No.	Application	Description
1.	Nanosensors	Used to detect and monitor contaminants, pathogens, and spoilage agents in food products
2.	Nanoemulsions	Improve the stability and shelf life of food products such as dressings, sauces, and beverages
3.	Nanocoatings	Provide protective barriers against bacteria, fungi, and oxygen to extend the shelf life of fruits, vegetables, and meat
4.	Nanofiltration	Used in water treatment processes to remove microorganisms and impurities from food and beverages
5.	Nanoencapsulation	Protect sensitive food ingredients like probiotics and vitamins from degradation during processing and storage
6.	Nanobiosensors	Enable rapid detection of foodborne pathogens and toxins through sensitive and specific bio-recognition processes
7.	Intelligent packaging	Monitor and control food quality parameters including temperature, humidity, composition of gas to ensure product quality and safety
8.	Nanopackaging	Incorporate antimicrobial agents or oxygen scavengers to preserve food freshness and prevent spoilage

Nanoparticles have demonstrated superior effectiveness compared to traditional encapsulation techniques in encapsulating and releasing substances. Nanoencapsulation provides protection for active ingredients against moisture, heat, chemical, or biological degradation during processing, storage, and usage. It also regulates the interactions between active ingredients and the food matrix, controlling the release of active ingredients at specific times and rates while maintaining compatibility with other compounds in the system. Additionally, these delivery methods can penetrate tissues deeply due to their small size, facilitating the targeted delivery of active substances throughout the body. To improve bioavailability and preserve active food components, various synthetic and natural polymer-based encapsulation delivery methods have been developed. Nanotechnology plays a vital role in enhancing multiple aspects of food products, including texture, appearance, flavor, nutritional content, and shelf life. Remarkably, nanotechnology has not only influenced these factors but has also introduced novel properties to food items.

Figure 8.1 illustrates various nanotechniques utilized for the encapsulation and delivery of functional ingredients.

Texture, flavor, and taste—Nanotechnology provides various solutions to enhance the quality and flavor of food, addressing factors such as appearance, texture, and taste. One widely used approach is nanoencapsulation, which improves the release and retention of flavors, leading to an enhanced culinary experience. For example, rutin, a dietary flavonoid with notable pharmacological activity, encounters challenges in the food industry due to its poor solubility. Nanotechnology offers a means to overcome this limitation and improve the solubility of rutin, thereby enabling its effective incorporation into food products. By utilizing nanoencapsulation techniques, the solubility of rutin can be enhanced, allowing for its successful integration into various food formulations. This not only improves the functionality and nutritional value of the food but also enhances the overall sensory experience for consumers. However, encapsulating rutin within ferritin nanocages improves its solubility, thermal stability, and resistance to UV radiation. Nanoemulsions, which are easily manufactured using natural food ingredients, are extensively used to deliver lipid-soluble bioactive compounds. They have the potential to increase water dispersion and improve the bioavailability of these compounds (Ozturk et al., 2015). Nanoparticles, due to their subcellular size, offer a promising approach to enhance the bioavailability of nutraceutical components compared to larger particles. This is because nanoparticles exhibit higher medication bioavailability and can release encapsulated compounds more efficiently and rapidly. In the food industry, various metallic oxides such as silicon dioxide (SiO_2) and titanium dioxide (TiO_2) have traditionally been used as coloring or flow agents in culinary products. Silicon dioxide (SiO_2) is particularly popular as a food nanomaterial for carrying tastes or scents in food products (Yang et al., 2015).

Nutritive value—Most bioactive substances are susceptible to the highly acidic environment and enzyme activity of the stomach and duodenum, including lipids, proteins, carbohydrates, and vitamins. Bioactive substances often face challenges in terms of poor water solubility and absorption in their non-capsulated form. However, encapsulation not only protects these substances from harsh conditions but also enhances their assimilation into food products. Researchers are developing small edible capsules made of nanoparticles to enhance the distribution of medications, vitamins, and delicate micronutrients in our daily diets. These capsules utilize techniques such as nanocomposite formation, nano-emulsification, and nanostructuration to improve the efficient distribution of nutrients such as protein and antioxidants. By encapsulating bioactive substances like flavonoids and vitamins within polymeric nanoparticles, the transport of these beneficial compounds to specific targets in the body is facilitated. This targeted delivery system allows for enhanced bioavailability and absorption of nutrients, leading to more effective nutritional and health benefits. The use of nanotechnology in encapsulating and delivering bioactive substances shows promise in improving the overall efficacy of dietary supplements and functional food products (Singh et al., 2017).

Shelf life—Nanoencapsulation of bioactive ingredients plays a vital role in extending the shelf life of food products by slowing down or preventing degradation processes until the product reaches its intended destination in functional foods. In functional foods, bioactive components are susceptible to degradation and loss of functionality due to the harsh environment. Nanoencapsulation provides a protective barrier around these components, preserving their functionality and preventing degradation. Additionally, edible nanocoatings applied to various food components act as barriers against moisture and gas exchange. These coatings can transmit tastes, colors, antioxidants, enzymes, anti-browning agents, and extend the shelf life of packaged meals even after the package has been opened. By modifying the characteristics of the interfacial layer surrounding functional components, chemical breakdown processes can be slowed down. For example, curcumin, which is the most active but least stable bioactive component of turmeric (Curcuma longa), tends to exhibit reduced antioxidant activity. However, when encapsulated, curcumin has been found to exhibit stability during pasteurization and under various ionic strengths. This demonstrates the protective effect of nanoencapsulation in preserving the stability and functionality of bioactive components, even under challenging conditions.

Overall, nanoencapsulation offers a promising approach to enhance the shelf life and preservation of food products, ensuring the delivery of bioactive ingredients in functional foods with maintained quality and efficacy (Sari et al., 2015).

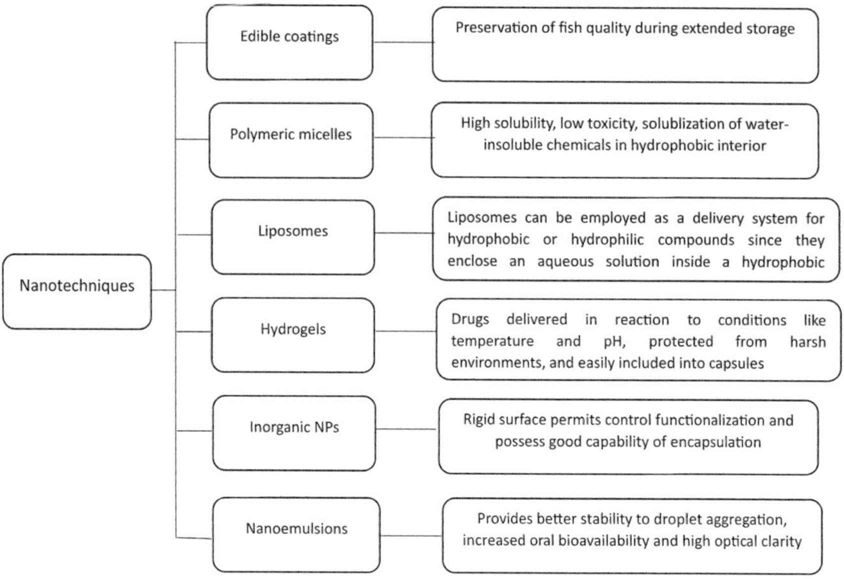

FIGURE 8.1 Different nanotechniques used for the encapsulation and delivery of functional ingredients.

8.3 NANOTECHNOLOGY AND FOOD PACKAGING

An ideal packaging material should possess qualities such as strength, biodegradability, and the ability to control gas and moisture permeability. Nanotechnology-based "smart" and "active" food packaging offers numerous advantages compared to traditional methods. It enables the development of improved packaging materials with enhanced mechanical strength, barrier properties, and antimicrobial films. Nanosensing capabilities also allow for the detection of pathogens, providing consumers with information about food safety. Researchers are exploring the incorporation of organic substances with antimicrobial properties, such as bacteriocins, organic acids, and essential oils, into polymeric matrices for antimicrobial packaging. However, these organic compounds are sensitive to physical conditions and cannot be used in food processing procedures involving high temperatures and pressures. In contrast, inorganic nanoparticles have demonstrated effective antibacterial activity at low concentrations and greater stability in harsh environments. This has led to increased interest in incorporating these nanoparticles into antimicrobial food packaging.

Antimicrobial packaging, categorized as active packaging, directly interacts with the food product or its surroundings to prevent or slow down microbial growth on food surfaces. Nanoparticles such as silver, copper, chitosan, titanium oxide, and zinc oxide have been found to possess antibacterial properties and are being investigated for their application in food packaging.

Nanocomposites and nanolaminates have found significant use in food packaging, providing protection against heat and mechanical damage, thus extending the shelf life of food products. Nanoparticles incorporated into packaging materials offer advantages beyond antimicrobial properties. They enable the production of high-quality food products with an extended shelf life. Polymer composites are created by incorporating various inorganic or organic fillers to develop packaging materials that are mechanically robust and resistant to high temperatures. The integration of nanoparticles into polymers has facilitated the production of durable packaging materials at a reasonable cost. Antimicrobial nanocomposite films, formed by embedding fillers with nanometric dimensions or nanoparticles into polymers, offer the dual benefits of structural integrity and barrier properties, as mentioned earlier (Singh et al., 2017).

8.4 NANOSENSORS FOR DETECTION AND MONITORING OF CONTAMINANTS

The two key components of human survival and well-being are food and the environment. These two components are intricately interwoven. The effects of these resources on people are widely known, and the problems brought on by the variations are also well known. The bulk of communicable diseases are thus most commonly spread through environmental elements like air and water. Due to their interrelated nature, these elements also influence food quality and hence, indirectly, human lives. Therefore, maintaining environmental and food quality factors is crucial. Numerous biological and chemical elements constantly infiltrate our ecosystems and can pose risks to human health upon contact. The industrial revolution has contributed to environmental degradation and the introduction of toxic anthropogenic substances into natural resources. Large quantities of untreated industrial and municipal wastewater are discharged into water bodies, and agricultural runoff often contains medications and pesticides, leading to a decline in water quality. Even when the initial concentrations of chemicals and wastes are within permissible limits, some substances have cumulative effects that can reach fatal levels over time. To minimize exposure and subsequent bioaccumulation, strict regulations are in place to manage these substances. Therefore, it has become imperative to implement food and water sensing systems for contamination monitoring, particularly in developing nations where wastewater management is currently inadequate. Nanosensors offer the capability to detect and sense various small molecules such as pesticides, pharmaceuticals (including medications and antibiotics), heavy metals, toxic organic compounds, volatile substances, as well as biologically active viruses, toxins, and bacteria. Extensive research and the contributions of experts worldwide have significantly advanced the field of diagnostics, food screening, and environmental monitoring of contaminants. The incorporation of highly effective, stable receptors has expanded the potential of various biosensing techniques, both in laboratory settings and in the field. These techniques have been created to explore the distinctive physicochemical features of nanomaterials. Also, with the development of nanotechnology, a wide range of accessible, quick, and targeted sensing methods have been created. Furthermore, we need powerful technologies to detect the existence of these pollutants and, occasionally, to measure them at their minuscule levels in order to maintain the quality of these parameters. Different nanosensors have been employed for chemical contaminants, detection of microbial contaminants, pesticides, metal contaminants, mycotoxins, etc. (Kannan and Guo, 2020).

8.4.1 NANOSENSORS FOR DETECTION OF MICROBIAL CONTAMINATION AND METAL CONTAMINANTS

The most frequent causative agents of a number of foodborne diseases are microorganisms such as bacteria, viruses, and protozoans. Colony-forming unit (CFU) counting is the traditional and culture-based method for detecting these microorganisms, but it is time-consuming, expensive, and requires expert labor (Yoo et al., 2007). There is an increased need for nonculture-based detection methods because not all types of microorganisms can be successfully cultivated in a lab environment. When detecting food contamination early on, nanosensors possessing higher sensitivity along with selectivity are preferable to colony-based counts because they can detect germs more accurately (Verma et al., 2015). The research is currently concentrating on the creation of methods that may be enhanced on portable sensing technologies to enable quick detection of a variety of infections by on-site analysis (Hossain et al., 2012). *Escherichia coli* O157:H7 bacteria, one of the most significant foodborne pathogens, is primarily responsible for contaminating various food sources, including raw milk, chicken products, cold sandwiches, vegetables, and drinking water supplies. This particular strain of *E. coli* poses a critical risk to food safety and can cause severe illnesses if consumed. Strict precautions and proper food handling practices are crucial in preventing the transmission and spread of this harmful bacterium (Bardasi et al., 2015). Jia et al. have developed a promising immunofiltration strip, which is a fast and efficient method for on-site detection of *E. coli* and other pathogenic bacteria linked to foodborne illnesses. This cost-effective alternative assay utilizes immunochromatographic test strips, making it highly suitable for routine surveillance purposes. The innovation offers a practical solution for quickly identifying and monitoring these bacteria, enhancing food safety protocols, and facilitating timely intervention measures (Jia et al., 2019). In order to identify *E. coli* O157:H7 with high sensitivity, quick turnaround, ease of use, and low cost, AuNPs (gold nanoparticles) photothermal action was used in an immunofiltration strip. This photothermal effect led to the amplification of the signal, thereby displaying the higher sensitivity of the strip. Also, in comparison to the conventional optical strip approach, the obtained sensitivity was approximately ten times higher (Kannan and Guo, 2020). Nanosensors serve as an indicator that responds to variations in environmental factors like temperature, humidity, microbial contamination, or item deterioration in storage rooms. It detects these changes and provides valuable information regarding the condition of the environment or the items being monitored.

Heavy metals including Pb, Cd, Cr, Ni, Hg, and As are responsible for causing carcinogenic effects to people at greater levels, even if consumed in trace amounts. Long-term consumption of food contaminated with heavy metals leads to significant illness, including neurological dysfunction and cognitive deficiencies. Numerous techniques, such as atomic absorption/emission spectroscopy and inductively coupled plasma mass spectroscopy, have been developed to detect heavy metal ions (Subramaniam et al., 2019). However, these conventional approaches call for either expensive equipment or labor-intensive preparation procedures. Consequently, recently developed nanomaterial-based biosensors provide a quick, affordable, and label-free method of effectively detecting hazardous heavy metals in food goods.

8.4.2 NANOSENSORS FOR DETECTION OF MYCOTOXINS

Mycotoxins, also known as toxins produced by fungi, are harmful compounds that can contaminate a variety of agricultural products. As a result, it is crucial to create new biosensing detection assays to identify mycotoxins (Granados-Chinchilla et al., 2018). According to statistical data, at least 25% of the grain products produced worldwide are ruined by mycotoxins. Mycotoxins can cause serious health issues such as vomiting, renal illness, liver disease, cancer, and even death when they are present in tiny amounts. Specifically, notable examples of harmful mycotoxins include ochratoxins, aflatoxins, trichothecenes, fumonisins, zearalenone, deoxynivalenol, ergot alkaloids, and tremorgenic toxins (Granados-Chinchilla et al., 2018). Naturally occurring aflatoxins are classified as Group 1 human carcinogens, while ochratoxins and fumonisins are categorized as Group 2B potential human carcinogens. Aflatoxin B1 (AFB1), considered one of the most hazardous mycotoxins, contaminates various agricultural food products such as tree nuts, peanuts, corn, grains, and animal feeds. The International Agency for Research on Cancer has designated AFB1 as a Group 1 carcinogen due to its significant health risks. Traditional detection techniques like HPLC, LCMS/MS, and ELISA are sensitive to AFB1 detection but require sophisticated equipment, labor-intensive processes, and complex sample preparation. Moreover, these methods are not suitable for real-time, on-site monitoring of AFB1 in samples. To overcome these limitations, researchers explored the feasibility of using a pH meter as a portable monitoring device, aiming to create a low cost sensor for AFB1 detection. The study employed an AFB1-responsive aptamer and cross-linked hydrogel to achieve precise AFB1 detection. When AFB1 binds to its aptamer with high affinity, the hydrogel network collapses, releasing urease into the solution. The released urease catalyzes the hydrolysis of urea, resulting in an increase in fluid pH. The developed portable sensing device demonstrated satisfactory results in detecting trace-level AFB1 content in food samples. This innovative approach offers a potential solution for rapid and cost-effective AFB1 detection, addressing the limitations of conventional methods and enabling on-site monitoring of AFB1 in food samples (Zhao et al., 2018). Due to its benefits of operational simplicity, high sensitivity, low cost, and portable on-field application, electrochemical biosensors are efficient and experiencing a great increase in attention (Hammond et al., 2016). For the quick and accurate detection of two mycotoxins, namely fumonisin B1 (FB1) and deoxynivalenol (DON), Gunasekaran et al. devised an electrochemical immunosensing technique (Lu et al., 2016). The created toxin sensor demonstrated good sensitivity and minimal matrix interference in real sample extracts from spiked corn samples, which is extremely helpful for applications involving the safety of food and feed products (Lu et al., 2016).

8.5 RISK ASSESSMENT AND REGULATIONS OF NANO-BASED FOOD PRODUCTS

Improper packaging performance can lead to the transfer of nanomaterial particles from the packaging into the food, posing a potential health risk when consumed. The severity of this risk depends on factors such as the toxicity of the nanomaterial used, the type of packaging material, the extent of migration, and the rate of food consumption. The evidence regarding health risks associated with inhalation of nanoparticles is steadily increasing. Misuse of nanotechnology-based products, bioaccumulation, and heightened activity can have adverse effects on health, raising concerns and risks related to safety and well-being (Jovanović, 2015). It is crucial to conduct a comprehensive risk assessment to examine any significant safety issues that may arise from human exposure to these substances at higher levels, particularly concerning potential long-term toxicity. This assessment will help evaluate and address the potential risks associated with prolonged exposure to these materials, ensuring the safety and well-being of the individuals involved. Certain studies suggest that silver nanoparticles present in packaging materials have the potential to migrate into food and subsequently be consumed by individuals. However, the extent of harm caused by such ingestion is still not well understood. The safety and potential risks associated with the consumption of silver nanoparticles from packaging materials require further investigation and research to provide a comprehensive understanding of their impact on human health. But in animals, these nanoparticles could build up in various organs like the kidneys, stomach, small intestine, liver, and spleen. Additionally, a single oral intake of ZnO nanoparticles may result in issues such as lung, renal, and liver damage. The use of titanium oxide and its final disposal can have an impact on both people and the environment, which raises the possibility of environmental and human health risks. Policymakers, concerned consumers, regulatory international bodies, and various other stakeholders are all paying attention to the application of nanotechnology. The interactions between nanoscale materials and their environment present challenges when it comes to tracking and monitoring their physical, chemical, and functional properties. These challenges arise due to the complex nature of how nanoscale materials interact with their surroundings. As a result, ensuring accurate traceability and monitoring of nanoscale materials requires careful consideration of these intricate interactions and their potential impact on the properties of the materials. This difficulty arises from the complex nature of how nanoscale materials interact with their surroundings. Therefore, ensuring accurate and comprehensive traceability and monitoring of nanoscale materials requires careful consideration of these intricate interactions and their potential effects on the properties of the materials. Without sufficient scientific study and consecutive case studies, drafting any rule or passing any legislation is quite challenging. Additionally, it is challenging since every nanomaterial responds differently to different products and processes. Although there has been very little development in the legislative process since it was initially proposed in 2003, a large number of food products based on nanotechnology are close to R&D trials and are waiting for legislation. However, some safety guidelines on the acceptable range or limits of nanomaterials employed in food applications must be established, as the above-mentioned behavior may change beyond a certain threshold. The use of nanoparticles in food packaging may be questioned by environmental authorities because it will ultimately be thrown away in some way. Few researchers have accumulated knowledge specifically related to laws on nanotechnology-based uses in food (Gergely et al., 2010; Hodge et al., 2014). There is a need for a global knowledge-sharing platform that brings together food processors, academics, researchers, and consumers to discuss and tackle various aspects related to the application, consumption, disposal, and long-term impacts of food. Such a platform would enable open debates, facilitate information exchange, and promote collaboration among stakeholders. By fostering dialogue and sharing expertise, this platform can contribute to a better understanding of the challenges and opportunities associated with food-related issues on a global scale. It will improve this promising field's research and applications even further. A general diagrammatic representation of components of risk assessment is provided in Figure 8.2.

8.6 CHALLENGES AND LIMITATIONS OF NANOTECHNOLOGY IN FOOD INDUSTRY

Despite numerous challenges, nanotechnology holds significant potential for creating innovative products and processes in the food industry. The key hurdle lies in developing edible delivery systems using cost-effective processing methods and formulating safe and effective compositions for human consumption. Ensuring the safety of food products necessitates preventing the migration and leaching of nanoparticles from packaging materials into the food. Occasionally, nanostructured materials are isolated due to migration from other sources. At the nanoscale, the behavior of materials differs significantly, and our technical understanding of their analysis is still limited. A comprehensive comprehension of nanoscale functionalities and toxicities is essential to enhance practical use and establish safety regulations.

Evaluating the effects of nanoparticles, potential risks, toxicological concerns, and environmental challenges is necessary. Nanoparticles have been observed to traverse biological barriers, penetrating cells and organs. Chemical processes involved in nanoparticle synthesis can generate harmful byproducts that contribute to environmental pollution. Therefore, when manufacturing, packaging, and consuming nano-based food products, comprehensive risk assessment programs, regulatory policies, biosafety measures, and public concerns should be considered alongside popularity and demand (Sahoo et al., 2021). Additionally, prior to commercial application and for the production of antibacterial nanoparticles using environmentally friendly materials, *in vitro* and *in vivo* investigations on nanoparticle interactions with living organisms are essential. Moreover, nanoparticles face limitations in widespread use in the food industry due to their high surface-to-volume ratio, which can lead to chemical reactivity, bio-reactivity, and potential harm to human health.

In addition, due to their high surface-to-volume ratio, nanoparticles face limitations in being widely used in the food industry due to their potential chemical reactivity, bio-reactivity, and potential harm to human health. However, this issue is not widely reported. Consumer awareness is crucial, and there should be transparency between consumers, farmers, and manufacturers to ensure that consumers are well-informed about the food they consume. Initially, people were fearful of Genetically Modified (GM) food due to genetic alterations, leading to bans in several countries (such as Russia and some EU countries banning cultivation but not import). In contrast, nanotechnology does not involve genetic alterations, making it a technology of the 21st century. The growing consumer preference for greener, eco-friendly, and chemical-free foods is exerting pressure on the food industry. Research indicates that people in developing countries consume around 10 trillion nanoparticles daily. However, the effects of nanoparticles on the genetic level are not well-studied. Therefore, it is crucial to conduct further research on the toxicity of nanoparticles. With smartphones being widely used globally, nanotechnology can facilitate food traceability from the farm to the dining table.

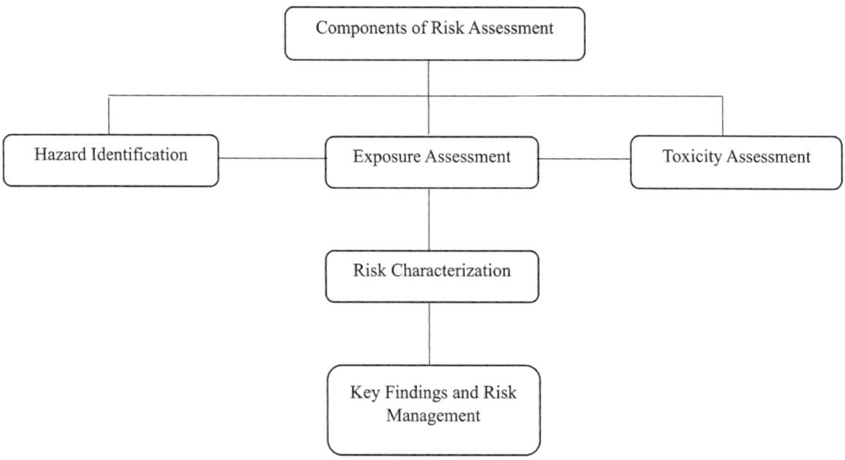

FIGURE 8.2 Diagrammatic representation of risk assessment of nano-based food products.

8.7 FUTURE PROSPECTS

In the upcoming years, the use of nanotechnology in food production is expected to expand the possibilities for creating and manufacturing functional foods. However, for this technology to dominate the food processing industry, specific rules and regulations need to be established to address the safety challenges it presents. Based on recent projections, nanotechnology is anticipated to emerge as a prominent advanced technology with a remarkable growth rate, expected to accelerate exponentially by the year 2050. This forecast highlights the potential for significant advancements and widespread adoption of nanotechnology across various industries and sectors, signaling its increasing importance in shaping the future of technological innovation. Its potential to provide solutions at both micro and macro levels positions it as a promising tool for solving various industrial and societal issues.

8.8 CONCLUSIONS

The utilization of nanotechnology offers several advantages in terms of food safety and quality. It positively impacts various stages of the food process, such as production, extended shelf life, nutrient preservation, quality control, and intelligent packaging. However, it is crucial to acknowledge that the application of nanotechnology also presents potential risks to human health, animals, and the environment. Recent studies suggest that nanoparticles may have toxic effects on biological systems, raising concerns about their safety. While some countries have already introduced nanotechnology-based materials in food contact materials (FCMs), it is vital to gather more data on the safety of engineered nanotechnology materials (ENMs) and nanoproducts. This will ensure appropriate regulation and safe application of these materials in FCMs. While the beneficial effects and potential uses of ENMs are well-documented, there is still a lack of knowledge regarding the potential ecotoxicological effects of nanoparticles. It is essential to address these gaps in understanding to make informed decisions and mitigate any potential risks associated with nanotechnology in the food industry.

ACKNOWLEDGMENTS

The authors would like to thank Mae Fah Luang University, Chiang Rai, Thailand for facilities and supports.

KEYWORDS

- **nanotechnology**
- **food safety**
- **nanopackaging**
- **nanocomposites**
- **contaminants**

REFERENCES

Bardasi, L., Taddei, R., Nocera, L., Ricchi, M., & Merialdi, G. (2015). Shiga toxin-producing *Escherichia coli* in meat and vegetable products in emilia romagna region, years 2012–2013. *Italian Journal of Food Safety, 4*. DOI: 10.4081/ijfs.2015.4511

Bratovcic, A., Odobasic, A., Catic, S., & Sestan, I. (2015). Application of polymer nanocomposite materials in food packaging. *Croatian Journal of Food Science and Technology, 7*, 86–94. DOI: 10.17508/CJFST.2015.7.2.06.

Chellaram, C., Murugaboopathi, G., John, A. A., Sivakumar, R., Ganesan, S., Krithika, S., & Priya, G. (2014). Significance of nanotechnology in food industry. *APCBEE Procedia, 8*, 109–113. DOI: 10.1016/j.apcbee.2014.03.010.

Dasgupta, N., Ranjan, S., Mundekkad, D., Ramalingam, C., Shanker, & R., Kumar, A. (2015). Nanotechnology in agro-food: From field to plate. *Food Research International, 69*, 381–400. DOI: 10.1016/j.foodres.2015.01.005.

Ezhilarasi, P. N., Karthik, P., Chhanwal, N., & Anandharamakrishnan, C. (2013). Nanoencapsulation techniques for food bioactive components: A review. *Food Bioprocess Technology* 2013, *6*, 628–647. DOI: 10.1007/s11947-012-0944-0.

Gergely, G., Wéber, F., Lukács, I., Illés, L., Tóth, A., Horváth, Z., Mihály, J., & Balázsi, C. (2010). Nano-hydroxyapatite preparation from biogenic raw materials. *Open Chemitry, 8*, 375–381. DOI: 10.2478/s11532-010-0004-4.

Granados-Chinchilla, F., Redondo-Solano, M., & Jaikel-Víquez, D. (2018). Mycotoxin contamination of beverages obtained from tropical crops. *Beverages, 4*, 83. DOI: 10.3390/beverages4040083.

Gupta, A., Eral, H. B., Hatton, T. A., & Doyle, P. S. (2016). Nanoemulsions: Formation, properties and applications. *Soft Matter, 12*, 2826–2841. DOI: 10.1039/C5SM02958A.

Hammond, J. L., Formisano, N., Estrela, P., Carrara, S., & Tkac, J. (2016). Electrochemical biosensors and nanobiosensors. *Essays Biochemistry,* 2016, *60*, 69–80, doi:10.1042/EBC20150008.

Hodge, G. A., Maynard, A. D., & Bowman, D. M. (2014). Nanotechnology: Rhetoric, risk and regulation. *Science and Public Policy* 2014, *41*, 1–14. DOI: 10.1093/scipol/sct029.

Hossain, S. M. Z., Ozimok, C., Sicard, C., Aguirre, S. D., Ali, M. M., Li, Y., & Brennan, J. D. (2012). Multiplexed paper test strip for quantitative bacterial detection. *Analytical and Bioanalytical Chemistry, 403*, 1567–1576. DOI: 10.1007/s00216-012-5975-x.

Jia, M., Liu, J., Zhang, J., & Zhang, H. (2019). An immunofiltration strip method based on the photothermal effect of gold nanoparticles for the detection of *Escherichia Coli* O157:H7. *Analyst 144*, 573–578. DOI: 10.1039/C8AN01004H.

Jovanović, B. (2015). Critical review of public health regulations of titanium dioxide, a human food additive. *Integrated Environmental Assessment and Management, 11*, 10–20. DOI: 10.1002/ieam.1571.

Kannan, P., & Guo, L. (2020). Nanosensors for Food Safety. In: *Nanosensors for smart cities* (pp. 339–354) Elsevier.

Lu, L., Seenivasan, R., Wang, Y.-C., Yu, J.-H., & Gunasekaran, S. (2016). An electrochemical immunosensor for rapid and sensitive detection of mycotoxins fumonisin B1 and deoxynivalenol. *Electrochimica Acta, 213*, 89–97. DOI: 10.1016/j.electacta.2016.07.096.

Ozturk, B., Argin, S., Ozilgen, M., & McClements, D. J. (2015). Formation and stabilization of nanoemulsion-based vitamin e delivery systems using natural biopolymers: whey protein isolate and gum Arabic. *Food Chemistry, 188*, 256–263. DOI: 10.1016/j.foodchem.2015.05.005.

Sahoo, M., Vishwakarma, S., Panigrahi, C., & Kumar, J. (2021). Nanotechnology: Current Applications and Future Scope in Food. *Food Frontiers. 2*, 3–22. DOI: 10.1002/fft2.58.

Sari, T. P., Mann, B., Kumar, R., Singh, R. R. B., Sharma, R., Bhardwaj, M., & Athira, S. (2015). Preparation and characterization of nanoemulsion encapsulating curcumin. *Food Hydrocolloids, 43*, 540–546. DOI: 10.1016/j.foodhyd.2014.07.011.

Singh, T., Shukla, S., Kumar, P., Wahla, V., Bajpai, V. K., & Rather, I. A. (2017). Application of nanotechnology in food science: Perception and overview. *Frontiers in Microbiology, 8*. DOI: 10.3389/fmicb.2017.01501.

Subramaniam, M. N., Goh, P. S., Lau, W. J., & Ismail, A. F. (2019). The roles of nanomaterials in conventional and emerging technologies for heavy metal removal: A state-of-the-art review. *Nanomaterials, 9*, 625. DOI: 10.3390/nano9040625.

Verma, M. S., Rogowski, J. L., Jones, L., & Gu, F. X. (2015). Colorimetric biosensing of pathogens using gold nanoparticles. *Biotechnology Advances, 33*, 666–680. DOI: 10.1016/j.biotechadv.2015.03.003.

Yang, R., Zhou, Z., Sun, G., Gao, Y., Xu, J., Strappe, P., Blanchard, C., Cheng, Y., & Ding, X. (2015). Synthesis of homogeneous protein-stabilized rutin nanodispersions by reversible assembly of soybean (Glycine Max) seed ferritin. *RSC Advances, 5*, 31533–31540. DOI: 10.1039/C5RA03542B.

Yoo, K. H., Lee, S. H., Kim, H.-J., Sung, K. W., Jung, H. L., Cho, E. J., Park, H. K., Kim, H. A., & Koo, H. H. (2007). The impact of post-thaw colony-forming units-granulocyte/macrophage on engraftment following unrelated cord blood transplantation in pediatric recipients. *Bone Marrow Transplant, 39*, 515–521. DOI: 10.1038/sj.bmt.1705629.

Zhao, M., Li, S., Zhou, L., Shen, Q., Zhu, H., & Zhu, X. (2018). Prognostic values of excision repair cross-complementing genes mRNA expression in ovarian cancer patients. *Life Science, 194*, 34–39. DOI: 10.1016/j.lfs.2017.12.018.

CHAPTER 9

Role of Nanotechnology in Sustainable Food Processing

PRIYANKA SUTHAR[1], PRIYANKA THAKUR[1], ANIKA[1], VIKAS KUMAR[2], ATUL DHIMAN[1], and SATISH KUMAR[1*]

[1]Department of Food Science and Technology, Dr. Y. S. Parmar University of Horticulture and Forestry, Nauni, Solan, Himachal Pradesh, India

[2]ICAR-Indian Institute of Millets Research, Rajendranagar, Hyderabad, Telangana, India

*Corresponding author

ABSTRACT

The modern-day approaches for healthy, functional, and convenient foods have shaped the food industry to adopt a continuous change in the product outfit to meet the ever-growing demands of consumers for innovative food products. In recent times nanotechnology has emerged as a novel method for the production, processing, and improvement in functionality, delivery of biomolecules, etc. taking advantage of the reconstructing nature at the nanoscale which beyond increasing the surface area also results in several other active benefits. The industry has shown its potential in important human and environmental health ramifications. Nano processing in the form of nanoemulsions and nanoparticles has revealed their potential in various fields of food processing and preservation including diverse fields like microbiology, processing, packaging, functional food, safety, and shelf life extension of food and/or food products. Nowadays, effective delivery of various bioactive compounds with the

Nanotechnological Advancements in the Food Industry.
Ishrat Majid, Bababode Adesegun Kehinde, and Sheikh Shafat Hussain (Eds.)

DOI: 10.1201/9781774919750-9

help of nanocarriers has been attempted in various food products with varying degrees of success though a bright future is expected keeping in view the high-end research efforts by scientists around the globe. The application of nanotechnology in various fields of food processing and preservation is anticipated as a significant tool in sustainable food processing for improving or maintaining the quality of food, restricting the spoilage, and thus reducing the waste generation and improving the nutritional and functional value of the foods. The chapter focuses on insight into the most significant applications of nanotechnology in food processing and also discusses the future researchable issues for most commercial applications of nanotechnology in sustainable food production and processing.

9.1 INTRODUCTION

The continuously growing population leaves the world with the ultimate challenge of establishing food security, safety, and the new global concern of sustainability. The scientific developments and novel discoveries have resulted in enhanced productivity of the farms while the postharvest losses still remain a challenge, especially in developing and underdeveloped countries. Even if food security is ensured the bigger challenge shall be ensuring nutritional security which beyond strengthening the human resources shall also reduce the burden of medical expenses and thus can ultimately lead to an improved lifestyle for the general masses. Despite the fact that global food production has increased several folds, the food supply chain seems to be highly insufficient and is often interrupted by very high spending on bio-resources for the production of food and its processing, high quantum of postharvest loss, limited value addition due to inadequate infrastructure and knowledge, inefficient value chain systems, and several other factors. Above all, the increased farm inputs for higher production have resulted in the adoption of innovative technologies through intensive and extensive agricultural practices. Further, novel technologies beyond an increase in production are also employed to reduce the volume of food wastage by effective processing, packaging, and distribution of fresh or value-added foods. Nanotechnology has emerged as a novel food processing adjunct that offers diverse benefits with novel value-added foods having improved properties. In the last decade, the food processing industry has become a multidisciplinary science-based activity with the involvement of experts of various domains with a collective and collaborative approach for healthy, functional, and sustainable food production, processing, and distribution. In the last decade, the role and applications of nanotechnology in food processing and preservation can be accessed from the IPRs and patents registered, nano material-based food production and processing trends, the up-growth of nutraceutical and functional food industry after the introduction of nanotechnological tools, improved distribution due to effective packing, safety, and quality of the food served, which shows the expanding features of various applications of nanotechnology in food science.

Applications of nanomaterials in food production and preparation are relatively new though enormous with important targets including effective, safe, and quality food delivery. In the face of the continuously changing food industry, nanotechnology has been seen as an effective tool to reshape the food ensuring smart delivery of nutrients with the highest bioavailability, chemical-free preservation of processed foods, enhanced and stable functional profile of the foods, rapid sampling for potential detection of various biological and chemical contaminants, encapsulation, enhanced delivery and bioactivity of various molecules, etc. (Neme et al., 2021). Nanomaterials can be obtained from different sources while the common feature, that is, increased surface area leads to infinite possible applications of such structures one of the recently established facts is that these materials perform exceptionally well in ensuring food security which has emerged as the prime factor to define the survival of any food industry. The creation of novel nanostructures from natural or synthetic materials enhances the physical and functional properties of food while nanoparticles are diversely applied for the protection of the food and the packaging material from chemical deterioration. Nanomaterials incorporation in the packing can assist in various functions like processing, preservation, prevention, and protection of the package flavor interactions during storage and thus can limit waste production. Nano-emulsion-based materials are seen as novel edible coatings and effective carriers for various antioxidants for quality retention and shelf life extension of fresh produce. Keeping in view the ever-changing demands of the food industry for creating consumer-friendly and need-based novel foods the introduction of nanomaterials in the sensors, smart distribution systems, coatings, intelligent packaging, capsules as antimicrobial agents for active packaging, etc. is anticipated in the near future. This chapter precisely focuses on the nanotechnological approaches in food processing and preservation for enhanced production, quality, safety, and distribution of the food.

9.2 NANOSTRUCTURES IN FOOD SYSTEM

Several foods include substances that are nanoscale in size naturally. Food proteins and other macronutrients, such as polysaccharides and lipids, for instance, have globular sizes between 10 nm and 100 nm and linear thicknesses of less than 1 nm. The structural components of nanomaterials often display different dimensions, including zero dimensions (quantum dots and nanoparticles), one dimension (nanotubes), two dimensions (nanofilms), and three dimensions (nanomaterials). Types of nanostructures with their food application are depicted in Figure 9.1. As mentioned, food systems naturally contain a range of nanostructured elements that comprise numerous forms of carbohydrates and lipids. These materials can be further used to develop systems like nanoemulsions and nanoencapsulation from natural nanomaterials (Bajpai et al., 2018). Casein and milk proteins, two examples of milk products, are examples of natural nanostructures. Liposomes, polymeric nanoparticles, and nanoemulsions are the most important synthetic nanostructured food systems. These nanostructures promote regulated release, increase bioavailability, increase solubility, and safeguard bioactive components during production and storage (Pathakoti et al., 2017). For the development and manufacturing of nanostructures with potential uses in the food sector, a number of techniques and materials have been used. Inorganic elements are one of the most researched materials for developing nanostructures with appropriate thermal as well as mechanical stability. In comparison to organic compounds, they may have improved magnetic, electrical, and/or redox capabilities depending on the chemical makeup, structure, size, and crystallinity. Silica, iron oxide (Fe_2O_3), noble metals, titanium and zinc oxides (TiO_2, ZnO), and iron constitute the most researched materials for the generation of nanostructures with the potential to be utilized in food systems (Massironi et al., 2020).

Because nanoparticles have more surface area per mass unit, they are believed to have more biological activity than larger molecules with the same chemical structure. This presents several food application possibilities. Nanotechnology can help to enhance the characteristics of bioactive chemicals by lowering the size of their particles, such as their transport capabilities, solubility, lengthened duration in the gastrointestinal system,

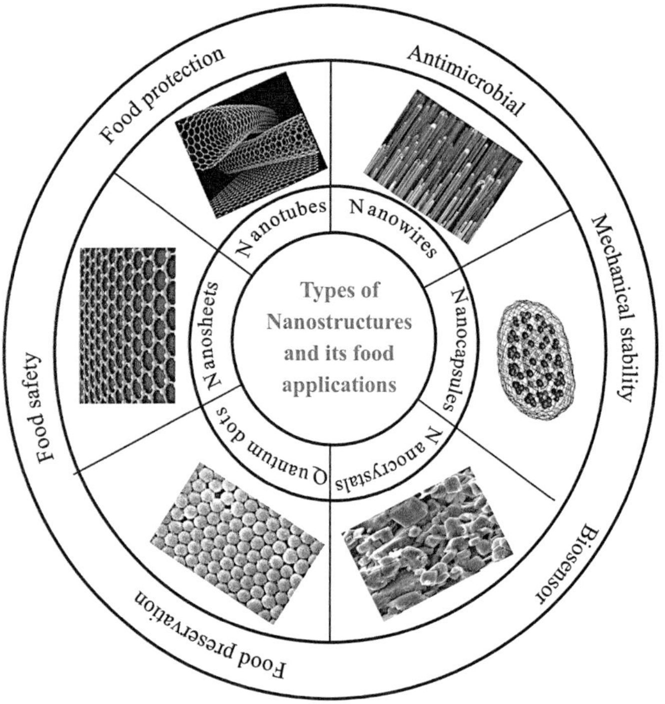

FIGURE 9.1 Types of nanostructures and their food applications.

and effective cell absorption (Sozer and Kokini, 2009). In food packaging technology, nanocomposites can significantly enhance the mechanical properties of biopolymers. Traditionally, inorganic fillers have been used to strengthen polymers in order to increase their mechanical strength and lower their cost. Conventional microfillers can be added to polymers, but doing so might have unfavorable side effects such as losing transparency, having the barrier performance suffer, and losing toughness (especially in film applications). Because of this, flexible food packaging applications utilize micro/nanoparticle-loaded polymers for reinforcement. Due to the aforementioned nano-effect, nanoparticles have the ability to boost mechanical performance (i.e., the reinforcing effect) without possibly impairing transparency or toughness. Additionally, the use of nanoparticles can result in an enhancement in the system's barrier performance. The primary factor for increasing the barrier performance of nanocomposites is the inclusion of high aspect ratio impermeable nanoparticles, which causes a significant reduction in the effective cross-section of the package while increasing the tortuosity of the diffusion channel. Recently, Campelo et al. (2020) compared both native and starch nanoparticles and reported that starch at the nanoscale has higher solubility, absorptive capacity, reaction surface, and biological penetration rate, hence can be used as a carrier to transport bioactive ingredients and mimic lipid micelles. In another study potential of biopolymer nanoparticles assembled from either plant (soy protein isolates) or animal (sodium caseinate or whey protein) sources were used to encapsulate norbixin and showed that protein nanoparticles enhance the stability as well as water solubility of norbixin (Liu et al., 2022). The weak water solubility, chemical instability, poor bioavailability, vulnerability to oxidation, deterioration during GI transit, and the food matrix impact limit the direct utilization of bioactive chemicals in foods. Due to these conditions, bioactive ingredient is less effective since it is not well distributed to the locations where its antioxidant activity is required. To increase the efficiency of using bioactives in meals, a number of methods have been used, including nanoemulsion, multi-layer nanoemulsion, colloidosome, solid lipid nanoparticles, and lipid nanocapsules. Since they may be easily included in packaging materials as coatings or by mixing in with the polymer's extrusion process, nanoparticles have shown to be an effective method for release control. The bioactive agent may be shielded by the encapsulation, which also gives it a strong foundation for easy inclusion into the packaging. Also, by adopting absorbing (scavenging) devices, undesirable substances like oxygen, carbon dioxide, ethylene, too much water, and so on may be eliminated. Using nanoclays to support scavenging systems is an easy way to create novel materials with a variety of uses (Cabedo and Gamez-Pérez, 2018).

9.3 ROLE OF NANOTECHNOLOGY FOR THE PRODUCTION OF FUNCTIONAL FOOD AND NUTRACEUTICALS

Despite concerns about consumer acceptance, safety and ecological effects, and regulatory issues, applications of nanotechnology to food have merely recently emerged. Nano-foods will bring about in a new age in food technology and have a significant impact on food sustainability, along with positive effects on human health and the environment. Nanotechnology has the potential to revolutionize the field of functional foods and nutraceuticals by enhancing their effectiveness, improving bioavailability, formulation enhancement, and enabling targeted delivery of nutrients and bioactive compound. Some application of nanotechnology in functional foods is provided in Table 9.1.

TABLE 9.1 Applications of Nanotechnology on Food Systems.

Sl. No.	Role	Applications	References
1.	Carrier material	• Encapsulation—nanoceuticals, nanocapsules, colloidosomes • Encapsulating bioactive antioxidants in electro-spun zein fibers	Pamidimukkala et al. (2023), Kasaai (2018)
2.	Food safety	• Nanosensors that integrate biotechnology, electronics, and nanotechnology have been developed for quick food detection • Nano-DNA barcode system is used to identify the genuine provenience of the product	Bai and Liu (2015)
3.	Functional ingredient	• Edible coatings made of gelatin incorporating cellulose nanocrystals	Fakhouri et al. (2014)
4.	Bioavailability	• Nanoliposomes have a larger interfacial surface area for contact with biological tissues, which may increase the bioavailability of drugs that are encapsulated within them	Singh (2016)
5.	Antimicrobial agent	• When padding solutions of different molecules on polyethylene/polypropylene (PE/PP) spun bond non-woven, nano-sized Ag colloids showed antibacterial capabilities against gram-positive and gram-negative microorganisms	Mikiciuk et al. (2017)
6.	Food packaging	• Metal-based nano sensors, nano barcodes • Nanosensors can also be used to determine the qualities of various foods, including wine, coffee, juice, and milk	Pamidimukkala et al. (2023), Thiruvengadam et al. (2018)
7.	Fortification	• Canola active oil fortified with nanocapsules containing tuna fish oil high in docosahexaenoic acid, an omega-3 fatty acid	Mousa and El-Hady (2018)
8.	Food additives	• Zinc oxide as a nanoparticle used as a food additive	Mohammad et al. (2022)

Enhanced Nutritional Delivery: Sensitive nutrients, such as vitamins, antioxidants, and omega-3 fatty acids, can be shielded against deterioration during processing and storage using nanoencapsulation techniques. These nutrients can be contained in nano-sized carriers, including liposomes or nanoparticles, which enables their regulated release and enhanced absorption in the body.

Improved Bioavailability: Numerous bioactive substances have low bioavailability, which means that their absorption and utilization by the body are poor. The solubility and stability of these chemicals can be improved by the use of nanoemulsion and nanocrystal technologies, increasing their bioavailability. This makes it possible for the body to utilize the advantageous elements included in functional foods.

Controlled Release Systems: With the help of controlled release systems made from nanoparticles, bioactive substances can be released gradually into the gastrointestinal tract. By ensuring that bioactive ingredients are exposed to the target areas in the body for an extended period of time, this regulated release can improve the effectiveness of functional foods.

Taste and Texture Enhancement: Functional meals can have their taste, texture, and sensory qualities improved by using nanoemulsions and nanocomposites. Nanoparticles can improve components' solubility and dispersibility, resulting in better sensory qualities and greater customer acceptance (Singh et al., 2017).

Safety and Quality Assurance: The food sector can make use of nanosensors and nanodevices for quality control and safety monitoring. These technologies can spot infections, pollutants, and spoiling signs, protecting the integrity and safety of functional food products.

Nutrient Fortification: Foods can be enriched with vital nutrients using nanotechnology. Nano-sized minerals, like calcium, iron, and zinc, can be added to food matrices to increase nutrient content without dramatically changing the final product's flavor or appearance (Mousa and El-Hady, 2018).

Numerous opportunities exist to improve the functionality and composition of nutraceuticals through the use of nanotechnology. These developments help to produce nutraceuticals that are more potent and efficient and have a favorable impact on people's health and wellness. Combined prescription drugs can be created using nanotechnology by combining many bioactive substances into a single nanosystem. By mixing complementary substances and enabling their simultaneous administration to the target site, this synergistic method can improve the therapeutic efficacy of nutraceuticals. To offer real-time monitoring of biomarkers, nutritional levels, or physiological factors, nanosensors and nanodevices can be incorporated into nutraceutical products. This makes it possible for individuals to receive personalized nutrition, track their health condition, and choose their nutraceuticals with knowledge (Thiruvengadam et al., 2018). Targeted distribution of nutraceuticals to particular cells or tissues within the body is made possible by nanotechnology. Lipophilic bioactive compounds can be delivered to specific sites using functionalized nanoparticles that can be made to recognize and interact with particular receptors or cellular markers. Liposomes, nanoparticles, and nanoemulsions are components of nanoscale delivery systems that enable encapsulation. These microscopic carriers shield sensitive minerals, phytochemicals, or probiotics from oxidation, stomach enzymes, or unfavorable environmental conditions. They can also increase their permeability, absorption, and release profiles, which will increase their effectiveness within the body (Singh, 2016).

It is important to keep in mind that the use of nanotechnology in functional meals is still in its infancy, and more investigation is required to solve any potential safety issues and guarantee regulatory compliance. To fully utilize nanotechnology's promise of enhancing human health and well-being, the food business must employ it in an ethical and responsible manner. Nano coatings and nanocapsules are two examples of how nanotechnology can be used in intelligent and active packaging systems to enhance various properties, including antimicrobial activity.

9.4 ROLE OF NANOTECHNOLOGY FOR HEALTHIER FOOD ADDITIVES

Nanotechnology has been long identified as a promising technology for maintaining food quality and safety in the food processing industry. The practice of nanotechnology in the food industry can be classified on the basis of usage as 'direct,' that is, the inclusion of nano-sized material directly in the food matrix in the form of food additives such as coloring/

flavoring agents, antioxidants, bioactive vitamins and minerals, nano preservatives, etc. and as 'indirect,' that is, combining the nanosize structure in the packaging system. The nanomaterials are also used in other forms such as gelation and viscosifying agent, nutrient delivery agent, and nanoencapsulation of tastes. The food ingredients with nanomaterial are being created with the promise of better consistency, taste, and texture. Currently, the nanomaterials are widely used as nanocarriers or as a delivery system for carrying food additives such as antimicrobial agents, that is, essential oils (thymol essential oil, cinnamon essential oil, clove essential oil, etc.), antioxidants and bioactive compounds (α-tocopherol, ascorbic acid, β-carotene, lipids- omega-3, or 6 fatty acids), without disturbing their basic morphology (Sahoo et al., 2021). The wider application of nanotechnology as a food additive can be conferred by the following major food applications.

9.4.1 NANOTECHNOLOGY IN TEXTURE ENHANCEMENT

Some nanomaterials such as nanostructured lipids have been incorporated in the food for its structural modification for instance in mayonnaise, ice creams, dairy spreads, etc. as a healthy alternative to the incorporation of fat-rich compounds mainly to simulate the creamy texture of the food. Nanotechnology as an appearance enhancer: various nanomaterials such as SiO_2, TiO_2, etc. have been conventionally used in the food industry as coloring agents or as a carrier of flavors and fragrances in the food. For instance, in gums, candies, cake icings, white sauces, etc. Nanotechnology for encapsulation of bioactive compounds: the nanostructured bioactive compounds such as lycopene, β-carotene, lutein, isoflavones, etc. are encapsulated in either oil-based or protein-based carriers to enhance the functional properties of the food. The nanoencapsulation enhances the bioavailability of these bioactive components and provides protection against the degradation of these compounds by oxygen or sunlight (Adeyeye and Fayemi, 2019).

9.4.2 NANOTECHNOLOGY IN NUTRIENT ENRICHMENT

According to Kulkarni et al. (2016), nanotechnology can help in masking the unpleasant taste or odor like fishy odor of omega-3 fatty acid by nanoencapsulation to enhance its consumption by the consumer and enrichment of various food products like cakes and muffins. Although nutrient encapsulation by nanotechnology has widely been implemented for various products, that is, β-carotene nanoencapsulation for beverages, calcium nanoencapsulation for flavored milk, biscuits, infant milk, chocolates, etc., the utilization of nanoformulations for the purpose of nutrient addition or fortification poses some challenges like vitamin fortification is challenged by its sensitivity to light, heat, pH, and oxygen. Mineral fortification also faces downsides, that is, iron fortification in the food can lead to the oxidation of lipids and vitamins resulting in rancidity and reduced bioavailability.

9.4.3 NANOMATERIALS AS ANTIMICROBIAL AGENTS

The nanomaterials incorporated in foods play a significant role in the prevention of various harmful and decay-causing pathogens by inhibiting their proliferation in the food commodity. For instance, nanomaterials, that is, heavy metals such as silver nanoparticles and nanocomposites, that is, Ag/LDPE, TiO_2, etc., have been recently used as non-traditional antimicrobial agents in the food industry. Silver nanoparticles which act as a source of Ag ions, bind to membrane proteins, form pits, trigger various morphological changes, and catalyze the production of ROS in bacterial cells, which ultimately causes oxidative stress and cell death (He and Hwang, 2016; Sahoo et al., 2021).

9.5 ROLE OF NANOTECHNOLOGY IN FOOD PACKAGING

Novel materials and packaging techniques have recently been thoroughly investigated and studied in a number of studies. The latest packaging trends involve nanotechnology-based nano reinforcement, nanocomposite smart or active, and naturally derived biobased matrix (Chausali et al., 2022). By acting as a barrier to stop oxygen and other spoilage-causing gases from scavenging the food, packaging serves as a main means of controlling spoilage and contamination as well as providing physical protection to safeguard food against temperature, external shock, vibration/ compression/ impact damage, and pathogenic infection. The ideal solution in this regard has been demonstrated by the use of nanotechnology in food packaging, specifically nanoparticles, and composites made of polymer-based materials (Mohammad et al., 2022). Various types of nanotechnology-based food packaging systems widely used in the food industry can be classified as:

9.5.1 ACTIVE PACKAGING SYSTEM

It is a purposefully created packaging system that incorporates elements that could absorb oxygen, release antimicrobials or antioxidants into or from packed food, or both. When active ingredients like nanoparticles, antimicrobial agents, and absorbers, that is, water vapor and oxygen absorbers are combined with a polymer in the packaging, the shelf life and quality of packaged foods are increased. Metal and metal oxide nanoparticles, that is, ZnO, Ag, and other nanoparticles have been incorporated into the packaging material for potential inhibition of the pathogenic strains (Mohammad et al., 2022). According to Esmailzadeh et al. (2016), ZnO-containing nanocomposites were developed for their action against *B. subtilis* and *E. aerogenes* and it was found to be effective against the growth of both *B. subtilis* and *E. aerogenes*.

9.5.2 INTELLIGENT PACKAGING SYSTEM

Intelligent packaging also popularly recognized as smart packaging is a technique that is used to monitor, find, and provide details about the environmental conditions and food quality inside the food package. The role of intelligent packaging can be classified into two main categories, that is, tracking the conditions and quality of food in real-time by identifying changes (such as color) and another one is its role as responsive packaging (smart packaging), which provides a source for releasing bioactive agents (antioxidants, antimicrobials, etc.) when food spoilage occurs. The nanomaterials have been widely studied by various researchers for their chemical and electro-optical properties and for their application in intelligent packaging as nanosensors for easy detection of gases, toxins, and harmful pathogens. Microbes, diseases, and associated toxins have all been monitored using fluorescent nanoparticles. A bio receptor found in nano biosensors that binds to analytes (microbes or related toxins) allows for the detection of toxins produced by numerous food spoilage bacteria.

For instance, aptamer-based biosensor consisting of poly(amidoamine) dendrimers as a nanomaterial and an Aflatoxin B_1 aptamer as a bioreceptor was used for toxin detection of a sample containing Aflatoxin $B_{1.}$

Nano coatings are thin layers of substances that are applied to the surface of products or packaging materials and are typically at the nanoscale level. These coatings can offer a variety of capabilities, including mechanical strength, barrier qualities, and even intelligent features. Nano coatings can be created to react to outside stimuli or to offer details about the packaged product in the context of intelligent packaging.

As an illustration, certain nanocoating can alter their color or transparency in response to changes in temperature, humidity, or the presence of particular gases or chemicals. These coatings can be used to visibly exhibit changes in color or transparency to indicate the freshness or deterioration of food goods. This can be especially helpful for perishable commodities because it makes it simple for customers and retailers to evaluate a product's quality and safety.

9.5.3 NANO CAPSULES AS ANTIMICROBIAL AGENTS IN ACTIVE PACKAGING

Drugs, perfumes, or antimicrobial agents can all be enclosed in nanocapsules, which are nanoscale containers or particles. Nanocapsules can be employed in the context of active packaging to release antimicrobial chemicals in a controlled way to increase the shelf life of perishable goods and prevent the growth of bacteria. The aluminosilicate clay known as cHNT, which is hollow nanotubular-shaped, is frequently utilized as a vehicle for active antioxidant and antimicrobial compounds. With the use of vacuum or freeze-induced loading procedures and actively agent-loaded HNT referred to as nano capsules, these agents are added inside porous HNT clay. As a result, HNT which has been packed with active substances will serve as a shield for extremely sensitive phytochemical and polyphenolic compounds found in antioxidants and antibacterial agents (Kumar et al., 2023). The release of antimicrobial agents can be triggered by particular circumstances, such as changes in temperature, pH, or moisture, by inserting antimicrobial nanocapsules into packaging materials or directly into the packaged product. The antimicrobial compounds are only released when needed according to this controlled release system, which maximizes their efficacy while reducing the possibility of interactions with the product or the environment and the antimicrobial activity of these nanocapsules can help inhibit the growth of bacteria.

Biobased packaging system: biopolymers incorporated with the nanomaterial have been widely used in the food industry as an alternative to plastic and non-biodegradable packaging material. The biobased packaging material is produced by either reinforcing it with nanofillers, that is, nano clay, layered silicate nanomaterials, etc. or by preparation of bio-nanocomposites. Nanoreinforcement is a novel technique in which the space between the polymer molecules is filled with nanofillers to enhance the strength, flexibility, and thermal or barrier properties of the matrix. Bionanocomposites such as chitosan, cellulose, and starch-based nanocomposites such as polylactic acid have been utilized widely as edible coating for fresh or processed commodities to enhance the shelf life of the commodity. For instance, Wang et al. (2022) developed the chitosan coating by incorporation of clove essential oil nanoemulsion to control *Burkholderia gladioli* and improve postharvest quality of fresh *Tremella fuciformis.*

9.6 CONCLUSION

The continuous search for more convenient measures to make food production, processing, distribution, etc. more effective and realistic has resulted in introduction of the nanomaterials in agriculture. Ever since their introduction the food industry has seen spectacular growth and emergence of new novel food materials with all new properties. The nanotechnological tools and their food applications are newly emerging developments in the last decade as reflected by the highly expanding adoption of the nanotechnological tools by various sectors of food processing and preservation, that is, production, processing, quality assurance, safety, value chain, etc. reveals the futures of the nanotechnology in food processing. Its fascinating phenomena have been revealed by various findings reported in the last decade, especially in the functional and nutraceutical food industry. The nanostructures show various improved properties due to their nano size, increased surface area, and highly catalytic nature while all these properties play a crucial role in food processing. Nanostructures are seen as potential tools for improving food processing, tailor-made foods with improved stability and sensorial properties, applications in sensing technologies for safe food processing, production and distribution, shelf life extension and quality retention during storage, minimized food losses, etc. Introduction of the nanostructures in packaging systems is also been seen as a research dimension to explore the miraculous potential of nano-material in improving the packaging. Above all, different nanoparticles used in food processing, packaging, and improvement are least studied for various health risks as the field is relatively new and is mostly studied for the beneficial effects associated with them. The nanostructures can easily penetrate through the cell wall and sometimes can show severe adverse effects upon consumption. The risks associated with nanoparticles or nanocomposites could be systematically studied for clinical documentation of any health effects associated with them. Further, future research shall focus on an intensive search for alternate, easy-to-use, user-friendly, and most importantly cheaper protocols creation of novel foods and degradation of the nanomaterial without any side effects.

KEYWORDS

- **food properties**
- **food safety**
- **nanostructures**
- **nanotechnology**
- **shelf life extension**

REFERENCES

Adeyeye, S. A. O., & Fayemi, O. E. (2019). Nanotechnology and food processing: Between innovations and consumer safety. *Journal of Culinary Science & Technology, 17*(5), 435–452.

Bai, H., Liu, X. (2015). September. Food nanotechnology and nano food safety. In: *2015 IEEE nanotechnology materials and devices conference (NMDC)* (pp. 1–4). IEEE.

Bajpai, V. K., Kamle, M., Shukla, S., Mahato, D. K., Chandra, P., Hwang, S. K., Kumar, P., Huh, Y. S., & Han, Y. K. (2018). Prospects of using nanotechnology for food preservation, safety, and security. *Journal of Food and Drug Analysis, 26*(4), 1201–1214.

Cabedo, L., & Gamez-Pérez, J. (2018). Inorganic-based nanostructures and their use in food packaging. In: *Nanomaterials for food packaging* (pp. 13–45). Elsevier.

Campelo, P. H., Sant'Ana, A. S., & Clerici, M. T. P. S. (2020). Starch nanoparticles: Production methods, structure, and properties for food applications. *Current Opinion in Food Science, 33*, 136–140.

Chausali, N., Saxena, J., & Prasad, R. (2022). Recent trends in nanotechnology applications of bio-based packaging. *Journal of Agriculture and Food Research, 7*, 100257

Esmailzadeh, H., Sangpour, P., Shahraz, F., Hejazi, J., & Khaksar, R. Effect of nanocomposite packaging containing ZnO on growth of *Bacillus subtilis* and *Enterobacter aerogenes*. *Materials Science and Engineering: C, 58*, 2016, 1058–1063.

Fakhouri, F. M., Casari, A. C. A., Mariano, M., Yamashita, F., Mei, L. I., Soldi, V., & Martelli, S. M. (2014). Effect of a gelatin-based edible coating containing cellulose nanocrystals (CNC) on the quality and nutrient retention of fresh strawberries during storage. In *IOP conference series: Materials science and engineering* (Vol. 64, No. 1, p. 012024). IOP Publishing.

He, X., & Hwang, H. M. (206). Nanotechnology in food science: Functionality, applicability, and safety assessment. *Journal of Food and Drug Analysis, 24*(4), 671–681.

Kasaai, M. R. (2018). Zein and zein-based nanomaterials for food and nutrition applications: A review. *Trends in Food Science & Technology, 79*, 184–197.

Kulkarni, A. S., Ghugre, P. S., & Udipi, S. A. Applications of nanotechnology in nutrition: Potential and safety issues. *Novel Approaches of Nanotechnology in Food,* 2016, 509–554.

Kumar, L., Deshmukh, R. K., Hakim, L., & Gaikwad, K. K. (2023). Halloysite nanotube as a functional material for active food packaging application: A review. *Food and Bioprocess Technology,* 1–14.

Liu, H., Zhang, Y., Zhang, J., Xiong, Y., Peng, S., McClements, D. J., Zou, L., Liang, R., & Liu, W. (2022). Utilization of protein nanoparticles to improve the dispersibility, stability, and functionality of a natural pigment: Norbixin. *Food Hydrocolloids, 124*, 107329.

Massironi, A., Morelli, A., Puppi, D., & Chiellini, F. (2020). Renewable polysaccharides micro/nanostructures for food and cosmetic applications. *Molecules, 25*(21), 4886.

Mikiciuk, J., Mikiciuk, E., & Szterk, A. (2017). Physico-chemical properties and inhibitory effects of commercial colloidal silver nanoparticles as potential antimicrobial agent in the food industry. *Journal of Food Processing and Preservation, 41*(2), e12793.

Mohammad, Z. H., Ahmad, F., Ibrahim, S. A., & Zaidi, S. (2022). Application of nanotechnology in different aspects of the food industry. *Discover Food, 2*(1), 12.

Mousa, R. M., & El-Hady, D. A. (2018). Nano-food technology and nutrition. *Environmental Nanotechnology, 1*, 59–74.

Neme, K., Nafady, A., Uddin, S., & Tola, Y. B. (2021). Application of nanotechnology in agriculture, postharvest loss reduction and food processing: Food security implication and challenges. *Heliyon, 7*(12), e08539.

Pamidimukkala, T., Irshath, A. A., & Rajan, A. P. (2023). An overview of nanotechnology used in the food industries. *International Journal of Research and Analytical Reviews, 10*(2), 595 https://doi.org/10.5281/zenodo.7939716

Pathakoti, K., Manubolu, M., & Hwang, H. M. (2017). Nanostructures: Current uses and future applications in food science. *Journal of Food and Drug Analysis, 25*(2), 245–253.

Sahoo, M., Vishwakarma, S., Panigrahi, C., & Kumar, J. (2021). Nanotechnology: Current applications and future scope in food. *Food Frontiers, 2*(1), 3–22.

Singh, H. (2016). Review article nanotechnology applications in functional foods. *Opportunities and Challenges, 21*, 1–8.

Singh, T., Shukla, S., Kumar, P., Wahla, V., Bajpai, V. K., & Rather, I. A. (2017). Application of nanotechnology in food science: Perception and overview. *Frontiers in Microbiology, 8*, 1501.

Sozer, N., & Kokini, J. L. (2009). Nanotechnology and its applications in the food sector. *Trends in Biotechnology, 27*(2), 82–89.

Thiruvengadam, M., Rajakumar, G., & Chung, I. M. (2018). Nanotechnology: Current uses and future applications in the food industry. *3 Biotech*, *8*, 1–13.

Wang, H., Ma, Y., Liu, L., Liu, Y., & Niu, X. (2022). Incorporation of clove essential oil nanoemulsion in chitosan coating to control Burkholderia gladioli and improve postharvest quality of fresh *Tremella fuciformis*. *LWT, 170*, 114059.

Index

For Product Safety Concerns and Information please contact our EU representative GPSR@taylorandfrancis.com Taylor & Francis Verlag GmbH, Kaufingerstraße 24, 80331 München, Germany